This intriguing series encompasses exciting trends and discoveries in areas of human exploration and progress: astronomy, anthropology, biology, physics, geology, medicine, health, genetics, and evolution. Sometimes controversial, these timely volumes present stimulating new points of view about our universe . . . and ourselves. Among the titles:

THE
HUMAN BODY
AND
WHY IT WORKS

Raymond L. Powis

A SPECTRUM BOOK

PRENTICE-HALL, INC., Englewood Cliffs, New Jersey 07632

Library of Congress Cataloging in Publication Data

Powis, Raymond L.
  The human body and why it works.

  (Frontiers of science)
  "A Spectrum Book."
  Includes index.
  1. Human physiology. I. Title. II. Series.
QP34.5.P68 1985   612   84-23791
ISBN 0-13-444969-X
ISBN 0-13-444944-4 (pbk.)

10   9   8   7   6   5   4   3   2   1

ISBN 0-13-444969-X
ISBN 0-13-444944-4 {PBK.}

Editorial/production supervision: Marlys Lehmann
Cover design © 1985 by Jeannette Jacobs
Manufacturing buyer: Frank Grieco

This book is available at a special discount when ordered in
bulk quantities. Contact Prentice-Hall, Inc., General
Publishing Division, Special Sales, Englewood Cliffs, N.J. 07632.

Prentice-Hall International, Inc., *London*
Prentice-Hall of Australia Pty. Limited, *Sydney*
Prentice-Hall Canada Inc., *Toronto*
Prentice-Hall Hispanoamericana, S.A., *Mexico*
Prentice-Hall of India Private Limited, *New Delhi*
Prentice-Hall of Japan, Inc., *Tokyo*
Prentice-Hall of Southeast Asia Pte. Ltd., *Singapore*
Whitehall Books Limited, *Wellington, New Zealand*
Editora Prentice-Hall do Brasil Ltda., *Rio de Janeiro*

CONTENTS

To my mother, Florence Mae,
who, when all academic doors were closed to me,
opened one door and found a way.

PREFACE

Few things can rival the human body in complexity and sophistication. Through the years, and especially during the last 50 years, research has produced a huge collection of information about how the body works. Most of this information, however, remains in research form— fragmented, specialized, and unrelated to how the whole organism works. This fragmentation of data is common to many current texts on physiology, leaving to the reader the business of putting it all together.

In contrast, *The Human Body and Why It Works* is not a detailed or comprehensive text on physiology but, rather, an attempt to look at physiology from a more global perspective. The vantage points of this "guided tour" or overview of physiology are six primary aspects of physiology:

1. The cell
2. Muscles and movement
3. Systemic circulation
4. Systemic regulation
5. Body interfaces to the outside
6. Adaptation

Beyond the singular value of gaining a more complete picture of physiological events, a global view of physiology offers a better understanding of disease processes and how contemporary pharmacology can be a source of effective therapy. And finally, a systemic view of physiology may help a student better use the excellent, more detailed texts on physiology that are now available. But whether or not a reader stops at this level or continues to more advanced material, the invitation to take a look within is open to all.

1

THE CELL AS PART
OF THE WHOLE

It is difficult to believe in something unless we can see it, which makes concepts like atoms, ions, molecules, and cells difficult to translate into a personal reality. Life is so whole and complete in most organisms that it is hard to think of life as being divisible into smaller parts, parts so small that they cannot be seen without special optical aids. Except in times of illness, the body's internal organs function with little conscious awareness on our part, and seldom if ever do we see ourselves or others as a composition of cells.

The desire to see cells created a set of technologies that give quite different views of the cell. Microscopes come in a wide range of forms, including optical, phase-contrast, electron (both flooding and scanning), and now ultrasonic [1]. But even with all these aids, life in the cell usually hides behind a set of static images, which is a little like trying to appreciate a football game through a set of still photographs. Understanding the meaning of the cell requires a vision of the cell as a mobile, dynamic expression of life.

Removing a few cells from a body and placing them under a light microscope would show very little of the cell clearly. The cell and its components interact poorly with visible light, forcing one to use more drastic measures to see them. The histologist who studies the micro-anatomy of organs and cells will look at cells by first removing them from the body (performing a biopsy), then killing and fixing them so they do not fall apart with further manipulation, mounting them on a slide, and then adding dyes of different color and affinity to mark the cell components. Under an optical microscope, the cells now appear red and blue, purple and black. They are still and unchanging, showing no more evidence of life than pebbles on a beach. Yet while they are alive these cells give us form and function, immunity, and body fluid control, regulating themselves, other cells, and whole organs. They communicate with one another, at times in a most detailed way, and through this communication and specialization they work in concert to bring us life.

The cell is a logical place to begin a discussion of physiology. The cell is the wellspring of life, and to understand how functions interact, we can look at the cell in health and in disease. The historical discovery process was just the opposite, however. Humans learned about diseases and discomforts long before understanding the underlying biology of the cell. To place the cell correctly in the scheme of things, we will start by dividing the cell into functional components called *organelles*. After looking at these organelles in particular, we will then re-collect them into the concept of a working, living cell, the basic unit of life. In the end, the whole process provides no complete definition of life, but we usually know life when we see it. Life outside the cell may be smaller (like the bacterium) or very specialized (like the virus), but the cell remains the smallest indivisible unit of physiological function.

From the viewpoint of the cell, physiology begins to look like a study of cells working in concert. The difficulties inherent in this approach become evident when considering what happens when we listen to music. We can hear a whole orchestra and not recognize all the contributing sounds that make up the total sound. When one of the sounds disappears, the orchestra may sound different, but it is hard to say just why. Physiology has a similar problem of pulling out the unique contributions to function made by all the working parts, some at a cellular level, others at an organic level. And often the changes brought about by disease or normal changes in body function can be as subtle as removing only a single instrument from an orchestra of several hundred.

So this chapter is about cells, and central to the discussion is an ever-recurring theme: Cells are alive. For each cellular process revealed that makes an organ work, we should not forget—cells are alive. In looking at specific cell functions, even the laws of thermodynamics will seem to be defied because—cells are alive.

Where do we look in the body for a cell? The answer is: everywhere. In fact, so many are present (about 100,000,000,000,000 at any one time), with so many different and unique facets, that using a single cell as an example for all is difficult if not impossible. Organs and tissues like muscle, brain, pancreas, kidney, skin, even bones and teeth are made up of or by living cells.

If we consider the anatomy of an "ideal" cell and compare it with real living cells, it is quickly apparent that living cells never look that ideal, except for that first cell formed by the fusion of sperm and ovum or a few of the early unspecialized daughter cells that emerge from this primary union. Complex, multicellular organisms are made up of cells that do not look alike, and cells that do not look alike often do not act alike. Shape and function are intimately connected. The first clue to a cell's function is often its shape and internal organization.

Along a single dimension, the largest cells in the body are nerve cells that must reach long distances between cells or from one organ to another. These distances can be up to a meter in very large animals. In contrast, one of the smallest cells is the red blood cell that travels through the vascular system, reaching all parts of the body. Its small size, about 8 microns (0.000008 meter or about 0.0003 inch) in diameter, makes this sort of travel easier for the human red blood cell. Smaller still are some of the white blood cells that travel freely through the body, often through walls of the vascular compartment. Shape and function are so intertwined that similar functions in different species produce look-alike cells, so that a histologist can, with some confidence, identify similar cells from a variety of living things.

Despite differences in cell shape, size, and activity, many of the structures inside a cell are common to nearly all. This is not a rigid rule, however. Histologists and cell biologists still have a problem knowing for sure that cell organelle "X" is indeed organelle "X" and not some variant form of organelle "Y". A few of these variant forms will appear in our discussions as we connect intracellular organization to extracellular whole organ or whole body functions.

An understanding of how the "ideal" cell works will require an examination of some of the more common organelles and how they contribute to cell function and cellular life. Our microscopic journey begins on the outside of the cell at the boundary of the cell's interior and exterior, the cell membrane.

THE CELL MEMBRANE

Many structures and molecules seem to just "float around" within the cell, unattached to anything. Yet they must physically or chemically work together to complete a particular cellular function. If these components need to be in close proximity to one another, one solution is to put some sort of membrane or sac around the whole aggregate to prevent components from wandering away from one another, which is a somewhat oversimplified view of the function of the cell membrane. On the other hand, a cell membrane is not just a mathematical boundary that divides inside from outside, or a passive piece of "cellulose" that only serves to limit how far molecules within the cell can wander. The cell membrane has shape, thickness, and qualities that identify it as an active cell component.

Avoiding some potential confusion will require a separation of two apparently similar yet different cell boundaries. A "cell wall" and a "cell membrane" are different, although they both surround a cell. *Cell wall* refers to the rather inflexible structure created by organisms such as bacteria. Conversely, *cell membrane* refers to the flexible, lipid-based, nonrigid boundary of the cell found in most multicellular organisms. This separation of concepts appears a bit early in the discussion, but this distinction will soon become valuable in understanding the cell membrane.

Forming the Cell Membrane

The cell membrane is made mostly of lipids, or fats, as they are more commonly called. Lipids do not dissolve in water, which makes the business of wall formation rather curious. Dumping the kinds of lipids normally found in cell membranes into water results in the spontaneous formation of cell membranelike structures. A look at the chemical nature of this response will not only help us understand how the cell membrane is formed, but also how the cell membrane can be a stable cellular component without continuously expending energy.

As just noted, fats do not dissolve in water. When lipids are added to water, they form small globules that minimize the number of fat molecules exposed to the water molecules. The grouping of lipid molecules together in this formation physically bonds the molecules together in close proximity. This is called *hydrophobic bonding* (*hydro* = water, *phobic* = fear, which makes these "water-fearing" bonds) [2]. In other words, the lipid molecules are bound together by their common aversion to water, simply because the lipids are insoluble in water. The idea of bonding atoms to one another usually conjures up ionic bonding as the first consideration, where the atoms are bound to one another with a strong electric field. Although hydrophobic bonds do not involve the chemical reactions that form compounds, the strength of a hydrophobic bond is still considerable, giving the cell membrane a great deal of stability.

But the lipids that appear in the membrane of a cell are not simply "water insoluble." Here the biology of the cell exploits the best of two worlds with molecules that have both hydrophobic and *hydrophilic* (*hydro* = water, *philic* = loving, "water-loving bonds") properties on the very same molecule [3]. And this is exactly the condition of cell membrane lipids. A portion of the membrane lipid molecule is hydrophobic, and that portion tries to stay out of water. At the same time, the hydrophilic portion "dissolves" into water, giving an unusual dual nature to the lipid molecule.

Chemical reactions often produce an uneven distribution of electrons in the resulting compound, forcing one portion or another of the molecule to have a net electric charge. Many of the physical and chemical properties of a compound come from this electric charge. With a net charge on one portion of the molecule, the molecule ends up with relative positive and negative portions. These compounds are called "polar." Water is the most widespread of these polar compounds.

The hydrophobic portion of a lipid molecule is uncharged, that is, it is electrically neutral. The hydrophilic portion of that molecule, however, carries a net electrical charge, which permits it to interact with the polar water molecules. This combination of hydrophobic and hydrophilic segments on the same molecule gives membrane lipids interesting and valuable properties. A schematic lipid molecule with hydrophobic and hydrophilic segments appears in Figure 1–1.

With one part "water-fearing" and the other "water-liking," membrane lipids will position themselves at a water–air interface, with their hydrophobic tails extending into the air (keeping away from the water), while the polar or hydrophilic ends extend into the water. The final arrangement looks like Figure 1–2.

Without a water interface, the molecules can take on an alternate organization called a *micelle*, which looks something like Figure 1–3. This is a common form for lipids in a water solution, and these globules appear in the blood returning from the gastrointestinal tract after a fatty meal [4]. All the polar ends point outward into the water, while the nonpolar ends cling together in the center of the micelle, avoiding the water.

Figure 1.1. *A schematic lipid molecule typically found in cell membranes. This molecule is really quite complex, despite its superficial simplicity. The charged segment (CH) includes phosphate and is hydrophilic. The lipid segment (LS) is a long, hydrophobic tail made of carbon (C) and hydrogen (H) atoms.*

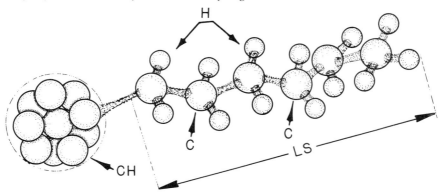

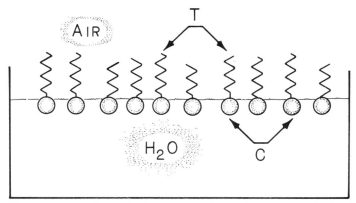

Figure 1.2. *Membrane lipids at a water–air interface. Because the lipid molecules have hydrophillic and hydrophobic segments, the molecules position themselves at the water–air interface with lipid tails in the air and charged heads (c) in the water.*

But the configuration we are most interested in looks like Figure 1–4. This new formation carries a number of advantages while still satisfying the general properties of lipids. The lipids collect to form a rather complicated lipid-membrane structure with the hydrophobic portions sandwiched inside the membrane (they are happily away from water). Now the formation separates the water into two volumes, one "inside" and the other "outside" the membrane. This double layer

Figure 1.3. *Lipid formation of a micelle in water. In a total water environment, the lipid molecules can form a micelle, where the charged heads (CH) point into the water and the hydrophobic tails (HT) congregate together. This formation is spontaneous and consumes no energy to hold its shape. These formations are often seen during fat absorption through the intestines into the bloodstream.*

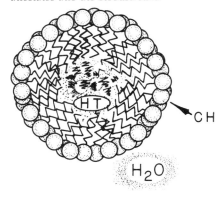

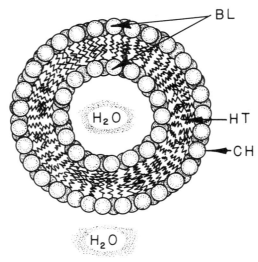

Figure 1.4. *Formation of a cell membrane bilayer. This is a variant form of the micelle with hydrophobic tails (HT) separated from the water. The charged heads (CH) face into the water on both sides of the bilayer (BL). Like the micelle, this formation is spontaneous and requires no energy to keep the arrangement stable.*

of lipids with hydrophobic bonds interior to the layer and polar ends extending into the water is the basic formation of a cell membrane.

The bilayer formation is spontaneous and represents the lowest energy level for the water-lipid system. Because this formation is spontaneous and sits at the lowest system energy level, the cell expends little or no energy keeping its membrane intact. For many cells, the double lipid layer is 75 to 100 angstroms (an angstrom is 10^{-10} meters) thick, which is less than 1/1000th of the red cell diameter. An electronmicrograph of a cell membrane appears in Figure 1–5.

What is the cellular payoff for using such a thin lipid layer for a boundary? Because the cell membrane consists of lipid molecules set in a water environment, the membrane spontaneously forms a very stable configuration. As a result, the cell maintains its integrity using the incompatability of molecules, an incompatibility that requires no energy expenditure to keep things intact. The hydrophobic bonds are not rigid, so the membrane remains together and yet is quite flexible, almost liquid. Still, a strong organization remains across the membrane thickness. These properties are those of a liquid crystal. Indeed, the lipid bilayer membrane is a *smectic* liquid crystal [5]. What a contradiction of thought—flexible crystals, but really no stranger than a liquid metal like mercury.

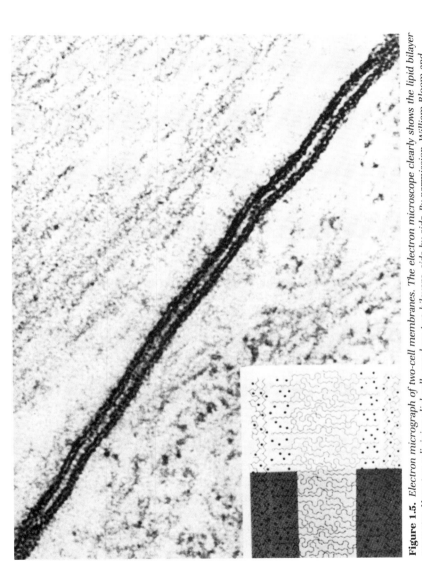

Figure 1.5. *Electron micrograph of two-cell membranes. The electron microscope clearly shows the lipid bilayer geometry. Here two adjoining glial cells produce two bilayers, side by side. By permission, William Bloom and Don W. Fawcett, A Text Book of Histology, Tenth Edition, W.B. Saunders Co., 1975.*

Lipids are the main constituent of the cell membrane, but a large variety of other molecules reside there too. These molecules give the membrane its more active roles in cell function. Covering the outside and inside the membrane, even penetrating through, are proteins. In addition, a number of cholesterol-based compounds reside in the membrane, although the function of these compounds is still unknown.

The proteins are quite central to cell membrane functions, and their activities immediately remove any notion that the membrane is only a passive sac to hold things together. And some membrane activities are independent of the cell's interior. For example, if we empty the cell's interior and replace it with saline, the membrane will continue to move specific ions across the membrane until the membrane runs out of energy [6]. It is time to take a closer look at some specific cell membrane activities.

Some Cell Membrane Functions

Much smaller than the cell itself are the ions, compounds, and molecules within the cell, and most internal cellular operations occur at this level of organization. The cell, therefore, can function properly only if the right ingredients are present, including energy. Along with the large and complex proteins, the cell also needs simpler ions such as sodium, potassium, chlorine, calcium, magnesium, and zinc. Not only do these ions need to be present, but in some cells they must be distributed properly on both sides of the membrane, which is a distribution largely controlled by the membrane [7]. This is a good example to look at more closely, because the characteristics of excitable cell membranes are fairly well understood.

We can get an idea of how a membrane can distribute critical ions by conducting a little mental experiment—taking a small piece of membrane and stretching it over a portal separating two fluid volumes. The concentrations of sodium on both sides of the membrane are initially the same. Figure 1–6 shows the arrangement of the experiment. Because the cell likes most sodium ions on the "outside," the first detectable thing the membrane does is move or "pump" sodium ions from one volume of fluid to the other. The net effect is to reduce the sodium concentration in one fluid volume and increase the sodium concentration in the other. As the membrane continues to move ions, it soon sets up a condition called a "concentration gradient," where one side of the membrane has a sodium ion concentration larger than the other. For a small, closed volume like the interior of a cell as shown in Figure 1–4, the membrane simply empties the interior of sodium ions in the process.

Setting up such a concentration difference is like rolling a large ball up a ramp. Moving the ball up the ramp requires doing work

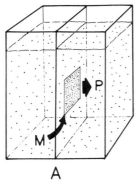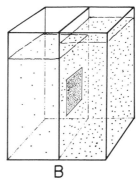

Figure 1.6. *Results of a membrane pumping experiment. A. The membrane M has a preferred pumping direction (P) for the ions, initially distributed equally on both sides of the membrane. B. The pumping action moves most of the ions to the right hand chamber, causing water to move with it (osmosis).*

against gravity. At the top of the ramp, some of this work can be recovered by letting the ball roll back down the ramp. The stored energy appears as kinetic energy. Keeping the ball at the top of the ramp, however, will require a constant equilibrium force.

And just as the ball on the ramp has a potential to move when released, molecules that have some degree of mobility and are gathered together will tend to move from regions of higher concentration to regions of lower concentration. Thus, the sodium ion outside the cell as part of a higher concentration also wants to "roll down" its concentration gradient. The cell, via the cell membrane, will expend energy to pump an ion through the cell wall against this concentration gradient. To keep the sodium out of the cell, the membrane then turns impermeable to the sodium ion [8]. Only the "pump" can get sodium ions through the membrane. They cannot travel back through the membrane to inside the cell. This keeps sodium ions moving in one direction, out of the cell. For now, it is of little concern where the energy comes from to push the sodium against its concentration gradient. The pump has an energy source which will be discussed later.

While excitable membranes like to keep sodium outside the cell, they also like to keep a relatively high concentration of potassium ions inside the cell. As the membrane methodically pumps sodium out of the cell, it just as vigorously pumps potassium ions into the cell. And the cell uses more than one sort of membrane pump to provide this separation of ions. For example, one ion pump works best when a sodium ion occupies the pump on the side of the membrane while a potassium ion occupies it on the other. The membrane then quickly exchanges the two ions through its thickness. The pump is appropriately called a sodium–potassium exchange pump [9].

The Cell Membrane   **11**

Despite a preferred direction for sodium and potassium ion travel, the membrane is not totally selective about what can diffuse through it. For instance, an excitable membrane is not completely impermeable to the sodium ion, as small numbers of sodium ions push through the membrane back into the cell because of the large concentration gradient between the inside and outside of the cell. Thus, the membrane is really only relatively impermeable to sodium, which has a hard time moving back into the cell's interior, but is not excluded entirely. On the other hand, the potassium ion passes more easily through the membrane down its concentration gradient, despite a vigorous pumping in the opposite direction.

Because potassium ions move through the membrane more freely than sodium ions, the potassium ions move out of the cell, leaving the interior of the cell slightly negative with respect to the outside of the cell. Thus, by selective permeability, that is, choosing which ion can move more freely through it, the membrane produces an electrical charge across its thickness. This membrane potential plays a central role in many physiological functions, and will appear in more detail later in this discussion.

Despite a diversity of cell shapes, varying preferences for ions and molecules, and a wide range of molecules attached to the inside and outside, all cell membranes have a similar construction, and from this common architecture comes a wide list of activities. For example, a membrane helps keep together the cell components that must work together closely; it puts up an effective screen to prevent the wrong ions and molecules from reaching the cell's interior chemical machinery; and it vigorously moves ions into and out of the cell's interior.

But the list is not finished. The membrane also provides a key function of "identity" within the body. The body's immune response must be able to separate cells that belong to it from foreign cells.

Posted on each membrane within the body is a chemical flag that says "this cell is part of this organism" [10]. An invading bacterium lacks this flag, and once the body determines it to be foreign, the body responds by killing and removing the bacterium. An individual with a transplanted organ faces another version of the same problem. The transplanted tissue has the wrong flag, and the body's immune system quickly works to destroy the new organ, even though the body may be unable to live without it. Problems can also arise if the right cells have the wrong flag, and once more the body's immune system begins to methodically destroy cells that may be quite normal, but wrongly identified. Such confusion within the body is called an autoimmune disease.

Along with providing an identity, the cell membrane also carries on a form of intercelluar communication observed in a phenomenon called *contact inhibition*. When normal cells grow and eventually fill an available space, cell division suddenly slows to a rate that only replaces dying cells. The contact of adjacent cell membranes seems to be the signal to the cells to stop further division within the whole cell colony. One conspicuous property of cancer cells is that they lack this contact inhibition, and despite the contact, continue to divide in a malignant fashion.

The list of membrane functions will end here with contact inhibition, although many more activities exist. The real limitation is the amount of time that can be devoted to this dynamic lipid layer. But the membrane activities essential to understanding organ functions are now in hand.

THE CYTOPLASM

Defining the Cytoplasm

In contrast to the cell membrane, the cytoplasm is not a cell structure but a region or volume within the cell. It includes the volume inside the cell, but outside the nucleus and other structures within the cell. If we imagine stepping inside a one-room cabin and looking around, the space we are interested in is inside the cabin walls but outside the stove and cupboards.

With only a light microscope to guide the way, the cytoplasm appears to be a clear, amorphous space, a site of chemical activity with little structural organization. But a strong spatial organization underlies this region as much as any other in the cell. Structuring the cytoplasm is an array of fine tubes and filaments that give the cell an effective skeleton to form around [11]. This recently discovered framework of ultrasmall elements gives the cytoplasm a structural complexity that rivals the skeleton of any larger animal. It was completely missed by histologists until the right technology revealed this internal anatomy.

The cytoplasm attracts attention not so much for its complex anatomy, but because many essential activities of the cell take place here. In addition, it holds a significant portion of the total body water. The water held in all cell cytoplasms (called intracellular water) adds up to 55 to 65 percent of all the body water in a healthy young man. *All* of the water in the body contributes about 60 percent of the total body weight in such an individual. Thus, all the intracellular water in all the body cells adds up to about 39 percent of the total body weight in a healthy adult male (12). The words "healthy," "young," and "male" are important modifiers because the amount of body water depends upon a person's state of health, age, and sex. A healthy young woman, for example, will have body water amounting to only about 50 percent of her body weight because of a larger proportion of body fat [12]. The water not in the cells occupies body spaces that fall under the general heading of "extracellular." The species of molecules and ions and the concentrations of these molecules and ions within the two water volumes, both intra- and extracellular, greatly influence cell life, and, as a consequence, body life.

The word "water" often produces a vision of a clear, free-flowing fluid that would pour from one container to another. But much of the water in the body, both inside and outside the cell, is not in a free-flowing state [13]. Instead, it exists in an organized, gel-like form with only small portions actually free to flow. The rest has the consistency of a very soft gelatin. As a result, most of the body water can easily change shape with little change in organization, and little actual water displacement.

This discovery about body water changed concepts about how to therapeutically manage body water and all the ions (electrolytes) essential to cell function [13]. Water in a gel suspension cannot rapidly flow out to replace free water lost from the body. The body cannot live without water and electrolytes, and people who have lost large amounts of body fluids through hemorrhage, extensive burns, or diseases such as cholera face death without some medical support. The extra- and intracellular gels have a major influence on how water shifts within the body when events disturb the normal water distribution. Knowing that the total body water is not free to flow not only changed the therapies for major losses of body water, but also explained some seemingly contradictory clinical observations [13].

Of all the molecules within the body, water is the most abundant and the most mobile. Water moves through nearly every cell membrane within the body with ease. The few exceptions to this rule appear as part of body water control in organs like the kidney and the gastrointestinal tract. Because water moves into and out of the cell so easily, some of the properties of water movement have to be recognized and controlled by the cell. One such primary process of water movement is called *osmosis*.

Osmosis and the Cell

To get an effective understanding of osmosis, we need to remember two basic principles: (1) Molecules like to move from regions of higher concentration to regions of lower concentration; and (2) Mixing a small amount of material A into liquid B increases the concentration of material A within liquid B, but at the same time *decreases* the concentration of liquid B (we now have fewer molecules of liquid B per unit volume). Let's carry these principles over now to follow events in a simple experiment.

The materials for this experiment consist of two column volumes of distilled water, separated at the lower end by a membrane with very small pores that penetrate through the membrane. The arrangement of parts is shown in Figure 1–7. The membrane is "semipermeable," that is, it lets water molecules pass through the pores freely, but it is impermeable to larger molecular structures. The final ingredient in the experiment is a salt such as sodium chloride. When sodium and chloride ions are in solution, they are too large to pass through the pores of the membrane. With these materials and conditions set up, on to the experiment.

With both columns filled with distilled water, the two columns will even out to the same height. Rapidly filling one side more than the other increases the hydrostatic pressure on that side and always leads to an equilibrium condition as the water moves from the higher pressure (taller) to the lower pressure (shorter) side. The water is free to move through the membrane, and equilibrium is always within reach.

Now, adding a small amount of salt to *one* of the water columns changes conditions on both sides of the membrane. The water begins to move *from* the column without salt to the column with salt. And this continues until the column of water with salt is higher than the column without. The process that is moving water through the semi-permeable membrane is called osmosis. Now let's find out what happened.

Putting the salt in the water obviously increased the concentration of salt within the water column. At the same time, however, the concentration of *water* within the column decreased (principle 2). Because the salt ions could not travel through the membrane to produce an equal concentration on both sides, the water moved from a higher concentration to a lower concentration (principle 1). The osmotic water-transfer stopped when the column of salt water was higher than the distilled water. This increase in height means an increase in pressure on that side, but in a direction opposite to the water flow. Thus, when the pressure in the salt-water column equalled the "pressure" produced by the imbalance in water concentration, water movement stopped, leaving an improbable situation of one column higher than the other, and conditions quite stable. Applying a pressure opposite and equal to the "pressure" produced by the difference in water concentration will stop the flow of water from one column to another. The amount of physical pressure required to just stop the fluid flow is an expression of the amount of energy difference between the two concentrations of water, which is called the osmotic pressure. The results of this experiment appear in Figure 1–7.

When we looked at a cell membrane earlier, a difference in concentration set up a charge across the membrane. Now a difference in

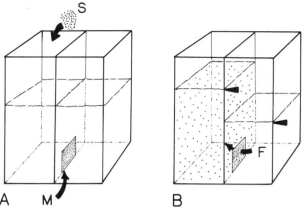

Figure 1.7. *Moving water with osmosis. A. The starting conditions are: pure water on both sides and a membrane (M) that can pass water but not the salt (S). B. Adding salt to the left side causes water to flow (F) from right to left until the difference in water heights (arrow heads) equals the osmotic pressure caused by the salt.*

concentration becomes a source of energy that moves water from higher concentrations to lower concentrations.

Let's now translate these conditions back to the cell. As stated earlier, the cell membrane is semipermeable, that is, it lets water go through quite easily, but excludes other molecules and ions from such free passage. And here arises a small dilemma for the cell. The membrane must not only determine what molecules and ions are present inside and outside the cell, but also keep the total concentration of water inside and outside the cell nearly the same. If the water concentration on either side of the cell membrane should differ, water will flow from its higher concentration to its lower concentration.

A cell that fails to maintain osmotic equilibrium (and thereby zero net water flow through its membrane) can rapidly get into trouble. If the water concentration outside the cell is greater than inside the cell, the cell will swell uncontrollably as water moves into the cell. If the swelling exceeds the elastic limits of the membrane, the cell bursts and dies. Many bacteria lean away from osmotic equilibrium by lowering their internal water concentration and surrounding themselves with a cell wall that is relatively strong and inflexible. As water moves into the bacterium, the cell wall prevents the bacterium from bursting, setting the pressure within the bacterium slightly above the outside [14]. Antibacterial agents that interfere with the formation of the cell wall lead to bacterial death when the bacterium lowers its internal water concentration only to find that the cell wall is unable to withstand the pressure, and both the cell wall and cell membrane burst, killing the bacterium.

If the internal water concentration were greater than the outside concentration, water would move from inside to outside the cell. Losing too much water leaves the cell unable to carry on life-sustaining operations, and it dies. For living things, the range of life-sustaining water concentrations in and out of the cell is quite narrow and requires rather careful management. It should be no surprise to find that the body has extensive controls to manage internal water and electrolytes, controls that are both sensitive and elaborate.

Most of the cell functions we have looked at so far happen because the cell has available energy. The major energy source for the cell is the *mitochondrion*, which is a small bacteria-like structure in the cell with an unusual anatomy and capabilities.

THE MITOCHONDRION

For a long time, mitochondria when viewed under a light microscope appeared to be nothing but mysterious rods and circles included inside many cells. The number of mitochondrial shapes seems to defy the principle of linkage between shape and function, but the shape common to all is just smaller than we can see, residing in the membrane of the mitochondrion. The different shapes confused scientists and led them to give a large number of names, nearly fifty, to the same cell organelle. Finding a common function removed the confusion, as biochemical examinations exposed the mitochondrial sameness.

Mitochondrial Anatomy

Mitochondria are a great deal smaller than the cells that carry them, with dimensions quite close to those of bacteria. Looking inside the mitochondrion requires more powerful tools than the light microscope. The electron microscope reveals a rather complex internal structure for the mitochondrion (Figure 1–8). For instance, the mitochondrion has two membranes. The outer membrane looks very much like a normal cell membrane, while the inner membrane has foldings called *cristae* [15]. On the foldings is chemical machinery that draws our current interest. The overall organization appears schematically in Figure 1–8.

Close views with the electron microscope show that the inner membrane appears to have polyps. These polyps, it turns out, are a highly organized set of molecules that are part of a complex energy transfer system. The energy comes from molecules that will give up

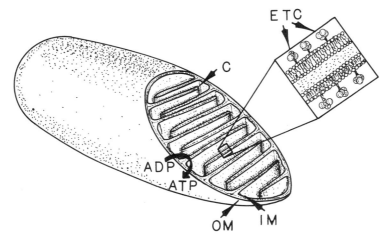

Figure 1.8. *Organization of a mitochondrion. Energy in the cell is chemically transported by ATP, produced by the movement of ADP into the mitochondrion. This organelle has two membranes, an outer (OM) and an inner (IM) membrane. The inside membrane folds to form christae (C). On the christae are the electron transport chain molecules (ETC).*

stored chemical energy, which the mitochondrion ultimately transfers to a special energy-carrying molecule called *adenosine triphosphate* (ATP). Although several methods of extracting energy are available to the cell, the most efficient consumes oxygen, releasing energy, carbon dioxide, and water along the way [16].

Making Useful Energy

The definition of a transducer is an entity that transforms energy in one form into energy of another form. The mitochondrion takes incoming energy in a form not useful to the cell and "transduces" that energy into a form that is useful, making the mitochondrion a biological transducer. A more dramatic example of this sort of transduction involves the chloroplast, another bacteria-like organelle found in plant cells that gives plant cells energy and a green color. The chloroplast transforms sunlight, a form of energy the cell cannot use directly, into chemical energy the cell can use directly [17].

The ability to make ATP is not the only curious property of mitochondria. They also have their own deoxyribonucleic acid or DNA. The DNA serves as the genetic template used in this case for constructing new mitochondria according to the cellular energy demands. If we step back two paces and look again, the mitochondrion begins to look like a bacterium. The mitochondrion has a specialized cell membrane; a variety of shapes; it makes a specific molecular product; consumes oxygen in making that product; and has its own DNA. We may never know just how this relationship between the cell and the mitochondrion started, but it all suggests an early cellular symbiosis (organisms carrying on functions in concert for mutual benefit) between a bacterium and a nucleated cell.

The detailed outline of energy conversion is too complex to go through in detail here, but the general principles of the mechanisms provide a glimpse of the driving forces behind cell function.

Within the mitochondrial matrix (the portion of the mitochondrion corresponding to the cell cytoplasm) are a number of enzymes that form a sequence of biological events called the *tricarboxcylic acid cycle* or *Kreb's cycle* [16]. The names come from the molecular structure the system dismantles to get energy, and the discoverer of the whole process, respectively. The word "cycle" describes how a set of active molecules enters into a chemical reaction with the "fuel" molecule and then appears complete and ready to do it again for the next fuel molecule. This chemical cycle removes energy from special organic compounds, and in the process, produces carbon dioxide and eight hydrogen ions for each swing around the cycle. The hydrogen ions now contain the energy extracted from the original compounds [16]. The hydrogen ions attach to a special molecular carrier, which delivers the energy-laden hydrogen to the inner mitochondrial polyps. The polyps are really a solid-state electron-transport chain. The hydrogen ions enter the chain, transfer the energy to another carrier (perhaps electrons) that pass down the chain in steps, producing ATP molecules at each step. At the very end of the transport process, oxygen combines with hydrogen to produce water. Notice that the oxygen we breathe ends up here within the mitochondrion and does *not* end up as the oxygen in the carbon dioxide we exhale with each breath. The carbon dioxide comes from steps in the Kreb's cycle, not the electron transport chain.

The details of energy extraction down the electron transport chain are not appropriate here, but we can follow a model to grasp the efficiency of the energy-extracting methods. Let's presume for a moment that a ball is sitting on a board extending out from the top of a ten-story building. The ball has potential to do work if it falls against a paddle. In addition, the kinetic energy acquired by falling one story against a paddle is enough to perform a desired amount of work or a "special task" at any floor level. If the ball were to fall straight to the ground from the tenth floor, it would have more than enough energy to do the special task. The energy "packet" we seek, however, is equal to the energy acquired by falling only one story, so, from all the energy gathered in falling all ten stories, only one story's worth is used; the rest is lost.

An alternate mechanism is to let the ball fall one floor, recover the energy, then fall another floor, again recover the energy, and so on until all the ball's potential energy is finally extracted at the ground floor. The whole process is more efficient because of a match between the amount of energy removed from the ball in each fall and the energy required to do the work. The electron transport chain works in this fashion, taking only enough energy at each step to chemically "charge up" an ATP molecule. The ATP molecule then moves to the various parts of the cell, carrying the energy needed for local activities.

Storing and Releasing Cellular Energy

Before leaving the mitochondrion, let's take a closer look at ATP and get some idea of how energy can be stored in and removed from this molecular structure.

As biological molecules go, ATP is relatively small and simple, with neither the size nor the outward sophistication of proteins or DNA. The ATP molecule is made of several smaller parts consisting of an adenine molecule and three phosphate groups hung on in a row. Thus, the "adenosine" for the adenine part and "triphosphate" for the three phosphates. But whatever the molecule lacks in outward complexity, it makes up for in internal sophistication. To function, ATP needs one more simple but essential component—magnesium. In the stereochemistry of ATP, a magnesium atom bonds to ATP in a specific form that permits stored energy to transfer from ATP to other molecules or processes. The chemical energy held within ATP resides in some complex way within the three phosphate bonds. Most of the stored energy transfers when the last ATP phosphate bond breaks, and ATP becomes ADP, *adenosine diphosphate* [18]. Excluding a few unusual situations, ATP handles nearly all the energy requirements of the cell.

The energy stored in ATP phosphate bonds is not free to indiscriminately transfer to any molecule or process within the cell. The transfer needs an intermediate that can link the ATP-energy to the biological process the energy is fueling. This cellular energy-transfer molecule is often a protein called an enzyme. The chain of events now looks like this: The mitochondria make cellular energy, transfer the energy to ATP, which transports the energy throughout the cell, and finally the energy links to a cell process through an enzyme.

Enzymes are biological catalysts that speed up chemical reactions in one direction or another [19]. They adhere to the basic rules of catalysts and although they speed up a reaction, they come out of the reaction just as they entered, capable of acting again. This renewing capability gives small numbers of enzymes rather large influences on chemical reactions. Enzymes are usually proteins made, in turn, of long complex chains of amino acids. The amino acid sequence specifies the physical way the whole molecule folds. The folding not only

THE CELL AS PART OF THE WHOLE

specifies the shape, but also the function of the whole molecule. An enzyme name comes from putting an "ase" suffix on the name of the affected molecule. Thus, an enzyme catalyzing ATP into ADP and energy is called an *ATPase*.

With an idea of where the cell's energy comes from, a good next question might be: How does the cell know what to do? To answer this question, we need to look at how the cell uses its master list of instructions, DNA.

THE NUCLEUS

Staining cells using current techniques and viewing them under a light microscope offers the vision of a single, round, darkly-stained structure holding position near the center of the cell (Figure 1–9). It is the cell's nucleus. Within the nucleus is a large amount of DNA *(deoxyribonucleic acid)*, some RNA *(ribonucleic acid)*, and a lot of proteins called *histones*. The stain is not uniform over the nucleus but mottled with light and dark regions. The more densely stained areas of the nucleus are collectively called *chromatin*. And often another darkly stained region appears in the nucleus, called the *nucleolus*. A difference in staining suggests some sort of chemical difference in the DNA itself

Figure 1.9. *A schematic cell with primary organelles. The basic organelles of the cell include: the nucleus (N) with nucleolus (NC); mitochondria (M); ribosomes (R); and lysosomes (L). The sarcoplasmic reticulum (SR) can be smooth or rough (RR) with granules formed by ribosomes. The SR communicates to the outside using a membrane that is continuous with the external cell membrane (CM).*

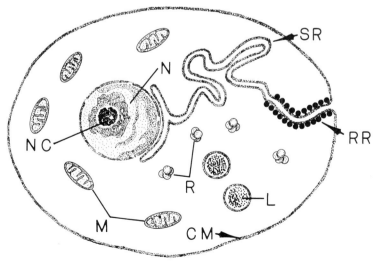

or in the proteins, called histones, bound closely to the DNA. We will come back to these differences in staining later in the discussion.

DNA Functional Groups

Nuclear DNA collects into distinct molecular aggregates called *chromosomes* (*chromo* = color, *somes* = bodies; they stain darkly, thereby becoming colored bodies). The chromosomes, however, only become visible just before cell division. Until then, they are packed away tightly in the cell nucleus, each visually inseparable from the others. At cell division, they become separated and identifiable, and a microscopist can separate them into groups according to the unique shapes they present under the microscope [20].

The chromosomes are small, but smaller still are genes, located on the chromosomes. How big is a gene? No one really knows for sure. And some current experimental results suggest that the gene is not as simple as we might think [21]. Still, the gene is much smaller than the chromosomes that are just visible under the light microscope.

Some extra emphasis on the relative size of a gene and a chromosome is necessary because it is an area of frequent confusion. Genes and chromosomes are casual parts of our conversation now, often without a clear distinction between the two. First, a chromosome is a very large set of folded DNA molecules or one very long folded molecule. Distributed along the length of DNA are genes, perhaps of variable length. The genes are right at the molecular level, and *cannot* be seen under a light microscope. They are simply too small. Chromosomes, however, are much larger and certainly are visible with a light microscope (Figure 1–10).

Because of the relative sizes of things, diseases that change either the shape of a chromosome or the number of chromosomes in a cell are detectable using a light microscope. On the other hand, genetic diseases offer no evidence visible with a light microscope. Genetic diseases appear indirectly when the cell expresses the malformed gene in a life process. An example of a chromosomal and a genetic disease should help keep things sorted out. Making three number 21 chromosomes in the fetus instead of the normal two will result in a condition called Down's syndrome [22]. These three chromosomes are visible under a microscope when they are specially stained and spread out on a microscope slide. This examination is called a *karyotype.* In contrast, sickle-cell anemia is caused by a single, wrongly coded gene that *cannot* be seen in a karyotype. It can be seen only when environmental conditions cause the red blood cells to shrink into the vessel-clogging sickle shape.

Other components of the nucleus are visible with light and electron microscopy. The electron microscope reveals that the nucleus is

THE CELL AS PART OF THE WHOLE

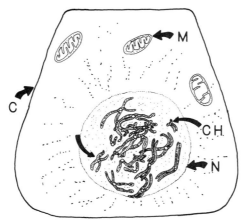

Figure 1.10. *Chromosomes within a cell nucleus. During mitosis, nuclear DNA replicates to form chromosomes (CH) within the nucleus (N). M are mitochrondria and the cell membrane is C.*

surrounded by a membrane very similar to the cell membrane. Under high magnification, holes (called *fenestrations*) appear to be distributed over the membrane surface. They may not be holes, but they appear as such under the electron microscope (Figure 1–11). Many investigators believe that through these fenestrations, messages pass from nucleus to cytoplasm, messages encoded on a type of RNA called messenger or M–RNA. Whether through fenestrations or the intact nuclear membrane, messages do travel from the nucleus to the cytoplasm, where they are interpreted into a wide range of cell functions.

Within the cell nucleus, in the form of long DNA strands, is one of the most compact memory systems in existence. Residing on the DNA molecules is a form of each of us, all in code. Each of us comes into physical existence when the early cells from conception correctly read the code and act on it. The DNA memory system was only a hypothesis until an experiment proved that in each intact nucleus of the body is a complete and accurate memory of the whole body.

DNA as a Memory

The experiment was really quite simple. A frog, which we will call Frog A, had a piece of skin removed from its gastrointestinal tract. A series of microdissections isolated a single cell and removed its nucleus. At the same time, a similar set of dissections removed the nucleus from a freshly fertilized frog egg and placed the Frog A nucleus into the fertilized egg cytoplasm. Now, if developmental instructions of the cell come only from the cell's nucleus, a twin identical to Frog A should

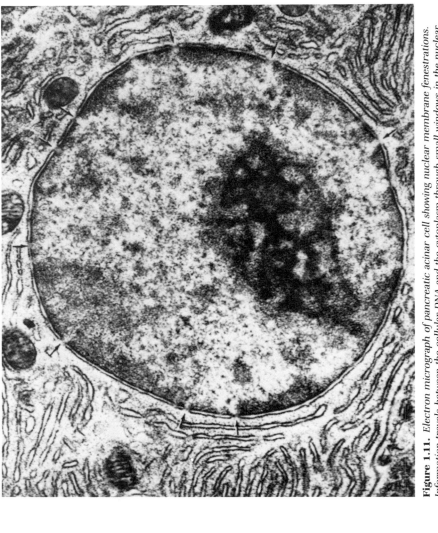

Figure 1.11. *Electron micrograph of pancreatic acinar cell showing nuclear membrane fenestrations. Information travels between the cellular DNA and the cytoplasm through small windows in the nuclear membrane called fenestrations (arrows). By permission, William Bloom and Don W. Fawcett, A Text Book of Histology, Tenth Edition, W.B. Saunders Co., 1975.*

grow from the fertilized egg. It did, identical in every way [23]. This process of making copies of an individual by using that individual's DNA is called *cloning*, a term now found in both fiction and nonfiction. And although the cloning of a human may be subject to both ethical and scientific debate, the memory of the individual contained on the DNA molecule is no longer a scientific issue.

Nuclear DNA is indeed a complete memory, and this memory resides in each intact nucleus. But the information for the whole organism is there as well, and a single cell, specialized and still part of the whole, uses only a small segment of the memory for its own instructions. This means that portions of the chromosomal DNA must be excluded and not available for message transcription. Could the chromatin seen in a stained nucleus be DNA in storage, or is it activated DNA ready for transcription? Many questions remain to be answered about the cell nucleus.

Given that the template of our cells is a code of ourselves—how does the code get interpreted to produce a person? Just what kind of information is in the code? The answers to these questions will require a lengthy sequence of steps and some imagination. The nuclear DNA molecule will be the starting point.

Using the Template

The search for the basic organization of DNA turned into a race after several near-solutions by researchers. The winners revealed DNA to be a double helix, each helix an exact compliment of the other [24]. And the first conclusion from this geometry is that the DNA memory system has a primary redundancy. Yet the whole system uses only four basic molecules, and these four make up the DNA coding for all living things. Restated, *the sequence of nucleic acids on DNA that specifies all living things uses only four nucleic acids to encode the information.* The sequence of nucleic acids specifies the code. What can such a code mean to the cell? Let's trace the information flow, starting at transcription, out to the cytoplasm and final expression, then collect things to form a coherent picture.

To control the sort of information that flows from the nucleus, the cell must protect its information from chance copies. The combination of histones (nuclear proteins) and a double helix protects the DNA codes from chance replication. To gain access to the DNA code, the double helix must be "unzipped," so to speak, so the sequence of nucleic acids is available for reading. The double helix is re-closed after the reading is finished [21]. A special enzyme opens the helix. When the helix is open, another enzyme called a *replicase* latches to the DNA and makes a strand of RNA coded exactly like the open DNA strand. This RNA is called messenger or M–RNA.

The M–RNA holding its message in a molecular sequence then moves to the cytoplasm of the cell through the nuclear membrane, while yet another enzyme closes the DNA helix. A second RNA called *ribosomal RNA* now enters the scene, waiting in the cytoplasm for the arrival of M–RNA. The ribosomal RNA (R–RNA) resides in small cell inclusions called ribosomes. A ribosome can attach directly to an M–RNA strand and assemble a protein according to the coded instructions on the strand of M–RNA.

A third RNA called *transfer* or *T–RNA* now enters the activities. T–RNA contributes to the protein synthesis by bringing amino acids to the ribosome that is moving slowly along the M–RNA strand, reading and accepting the proper T–RNA–amino acid complex to assemble the final protein molecule. Each of the twenty amino acids used to construct proteins has a separate and unique T–RNA molecule. At the end of assembly, the ribosome releases the final product, an assembled protein molecule ready to perform its cellular or systemic task. A diagram of this replication sequence appears in Figure 1–12. Right now, it seems a very long way indeed from that protein molecule to the body seen in a mirror. The intellectual journey from DNA to protein

Figure 1.12. *DNA-dependent protein synthesis. The cell's memory is a collection of DNA molecules in the nucleus of the cell, surrounded by the nuclear membrane (NM). Nucleic acids (NA) enter the nucleus to form messenger RNA (mRNA), which enters the cytoplasm to function as a template. Transfer RNA (tRNA) brings amino acids (AA) to a ribosome (RB) reading the template to form the protein sequence (P).*

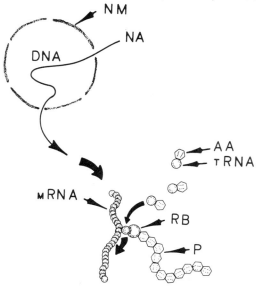

THE CELL AS PART OF THE WHOLE

to body, however, is not quite so far when we consider the potential role of any synthesized protein molecule in the body.

The link between the code in the nucleus and cellular functions is an active protein. Nearly everything the cell does is mediated through a protein of one sort or another. If a cellular function does not happen with a protein, then a protein is used to make or control the molecule that does bring about the cellular event. DNA is coded for proteins, not "life" as we might think of it. Once a cell's functions are defined by the proteins present within it, the collection of cells into organs specifies an organ's functions. Organs, on the other hand, support life in the whole organism. Instructions that radiate from the nucleus tell cells when to divide, when to specialize, what to specialize into, and which cells to collect together for a common function.

The instructions even reach the nervous system, telling the whole organism what to do with a particular stimulus (what we call instincts), and yet keep the parts working together as a single functioning unit. Thus, the flow of information begins at the headwaters of nucleus and DNA, and flows to the cell through RNA and proteins. The cells collect into organs that deliver metabolic fuel, pump oxygen to the cells, and remove wastes, all contributing an ability to function as a unified whole. Through this flow of information, the sequence on a strand of DNA we cannot see or feel codifies all of life.

Although the sphere of influence for nuclear information can extend from the cell to the whole body, the basic unit of DNA control is still the cell. For most tissues, investigators observe a nearly constant ratio between the cell volume and the amount of DNA in the cell [25]. The limits on the ratio are defined either by the distances required for nuclear information to reach the more distant corners of the cytoplasm or by the number of activities pursued by the cell. As a result, the ratio of cell volume to DNA content is generally large for cells that are mostly quiet, like supporting cells, and small for very busy cells like those in the liver.

Some organs respond to increased functional demands by increasing in size. This sort of growth can result from either an increase in the number of cells (called *hyperplasia*) or an increase in cell constituents, increasing the cell size (called *hypertrophy*). Hypertrophy in some tissues includes not only an increase in cell size, but also an increase in the amount of DNA in the nucleus [26]. The DNA increases are in even amounts, that is, two or four times the DNA content. The additional DNA comes from DNA replication, and because a double helix is the physical configuration of DNA, it grows by multiples of two. When a cell has more than the normal amount of DNA in its nucleus, it has a condition called *polyploidy*. For example, twice the normal DNA is a diploid nucleus; four times is a tetraploid nucleus. Polyploidy is more frequently seen in animals that, like humans, have a terminal adult growth pattern. Polyploidy is not frequently seen in animals like rats that continue to grow throughout an adult life. Although a connection between the cell volume and the amount of DNA seems solid, the reasons for polyploidy and its effects on the efficiency of cellular control are not well understood.

Following instructions from the nucleus is another cell component called the endoplasmic reticulum.

THE ENDOPLASMIC RETICULUM

Many cells contribute to body function by manufacturing a material for secretion that can communicate to other cells, protect other cells, or carry on a wide range of extracellular functions. This manufacturing capability is most evident in cells that make mucus to line passages and cavities and cells from the liver. In these cells, the function is quite obvious, but it becomes more subtle in cells less studied or less dramatic in what they do.

Endoplasmic reticulum (ER) appears in cells in proportion to the amount of molecular synthesis occurring within the cell. Thus, cells that support and hold organs together generally have very little ER, whereas the cells that secrete specialized materials such as special enzymes, hormones, or mucus, have large amounts of cellular ER. All the functions of ER are not completely known, and confusing things further are the highly specialized forms of ER we find in cells like skeletal muscle that do not secrete a material but sequester an ion.

Common to the many forms of ER is a communications link between the nucleus of the cell and the extracellular fluid. The communications link has an appearance like the diagram in Figure 1–9. For the cell, this link to the outside rapidly moves secretions from the cell to the outside.

Many active cells have ER membranes with small nodules of ribosomal RNA scattered along the surface of the membrane, giving the membrane a granular appearance under the light microscope. This is called *granulated* (or rough) *endoplasmic reticulum*. Other forms of ER appear without these granules, and the ER membrane in these cells looks smooth. Its name is *smooth endoplasmic reticulum*.

For most cells that carry on a primary role of secreting some kind of substance, the ER receives instructions for making the secretions from the cell nucleus. In these cells, the communications link between nuclear DNA and the ER seems almost continuous. ER in skeletal muscle, on the other hand, is smooth and specialized to sequester and release calcium ions as part of the controls for muscle contraction [27]. This specialization for ER deserves a new name and is called *sarcoplasmic reticulum*.

Beyond the cell membrane, mitochondria, nucleus, and endoplasmic reticulum, we have several other cell components yet to consider, some with functions we understand, and others that we still label "function unknown." Because many of the cell components seem to have only a specific use, we will introduce them as they appear naturally in the discussion of a particular physiology.

Central to everyday life is an ability to replace cells that die or are lost through normal interaction with the world outside the body. We normally lose cells at an enormous rate, and recovering from these losses requires making new cells.

MAKING A CELL

We all begin life as a single cell made by the fusion of sperm and ovum. From this single cell, we grow to the complicated organism that presents itself to a physician when the parts are not working well. Getting to an adult from a single cell means making new cells, and new cells means cell division.

In general, the cells of a fetus and the cells of an adult are nearly the same size. As the body grows, increasing the size of organs and limbs, the body simply makes more cells of each type to increase the size of everything. The skin protects both along the gastrointestinal tract and on the outer surface of the body and loses a large number of cells in the process, and all have to be replaced to maintain body integrity. New cells can come only from cells that are already alive and capable of division. This simple truth replaces much older notions like spontaneous generation, which suggested that old rags could produce animals such as mice. Despite the enormous number of cell divisions that occur in the body, some cells are transformed into specialized entities that never again divide, and when these cells die because of disease or age, they are never replaced. The result is a body made of a blend of cells, ranging from rapid turnover cells with a very short life span to cells that never change over a complete life span.

The rules governing this process called *differentiation* (a cell becoming specialized to carry out a single role in the body) are not well understood. But the process of starting with a few ordinary cells in gestation and ending up with organs like muscle, skin, brain, and liver, each with cells that are unique in shape and function, is expressed in each child born. We will spend some time looking at the process of gestation and birth later, in Chapter 6.

Without the advantages of modern technology to help them investigate events within the cell and its range of activities, early biologists would look at cells under a light microscope and see nothing happening. But once in a while, chance would work in an observer's favor and something new would appear in the collection of cells. A cell would be caught in the act of dividing. After many observers and observations, it eventually became clear that cells were obviously either in division or at rest, with little evidence of anything else happening. These early cell-watchers quite reasonably divided cell activities into a resting phase and a division phase, which they named *mitosis*.

For growth and body maintenance, cell division is the beginning and the end. And cell divisions are not random. It appears that a cell goes through a well-programmed cycle, extending from one cell division to the next [28]. The length of this cycle can range from as little as 10 hours for fast-dividing cells, to several thousand hours for others. And for some cells, the cycle extends over the lifetime of the whole organism. Nevertheless, the same components of the cycle appear for any sort of cell division, regardless of the cycle length. Even cancer and tumor cells will show the same cycle, but with an unnatural rhythm.

Getting Ready for Division

Cell biologists divide the cell cycle into two segments or phases because these two phases have distinct visual and biochemical characteristics. The first unambiguous phase is the *D* or *division phase* [29]. During this time, the cell undergoes division (mitosis), and the visible changes within the cell leave no doubt about what the cell is doing. The physical process of division, which we will look at in more detail soon, is unmistakable under the microscope.

The next distinctive segment of cell division is the so called *S* or *synthesis phase* [29]. During this period of time, the cell replicates DNA in preparation for the division soon to come. The process of making more DNA is invisible to the eye and requires biochemical techniques in order to be "seen." Between these two defined time-segments are gaps that carry no real distinctive, identifiable activities. The gap between the end of the D phase and the start of the S phase is called the *G1* (the G is for gap) *phase*. From the end of the S phase to the start of the D phase is the second gap called *G2*. Now, including the gaps, S, and D phases, the cell cycle divides into four major segments in a sequence of G1, S, G2, then D. When the cell rests between divisions, it does so in G1. The total cycle of events is shown in Figure 1–13.

The G1 phase is the resting phase for the cell. During this time, however, the cell carries out its functions in support of organs or the whole body, so "rest" is the wrong word here. During the S (synthesis) and the D (division) phases, the cellular DNA is not available for DNA transcription, which leaves a small segment, G2, and a larger segment, G1, as the only times available for DNA-dependent protein synthesis.

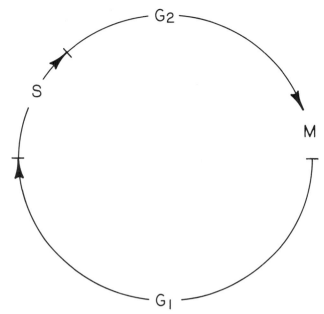

Figure 1.13. *The basic cell cycle. The cell goes through a regular cycle that leads to replication. Most of the cell's life is spent in G1, the resting phase. The cycle begins with the S or DNA synthesis phase. The cell then enters the G2 phase for a short time, then moves to mitosis M. After mitosis, both daughter cells again enter the G1 resting phase.*

The G2 segment, however, follows right on the heels of new DNA synthesis, and any DNA transcription at this time runs the risk of producing too much information because the cell has a double dose of DNA. This leaves G1 as the only logical choice (this is a little like the reasoning of Sherlock Holmes—eliminate the impossible and the remainder must be true). G1, therefore, is logically the start point and the end point for the cell cycle.

Prior to making two cells out of one, the nuclear DNA must be copied to provide DNA for the two daughter cells, which occurs during the S phase of the cell's cycle. During this phase, the DNA double helix must be unwound and reproduced by a replicase. But the original DNA is not rejoined. Instead, the new copy of the old DNA joins with the old DNA. As a result, the DNA distributes in a manner called *semiconservative*, that is, each new double helix comes from the combination of one old DNA molecule and one new DNA molecule [30]. Each of the two new daughter cells gets an equal share of old and new DNA (see Figure 1–14). With this distribution method, the dividing cell keeps half of its old DNA and donates the remainder to the new

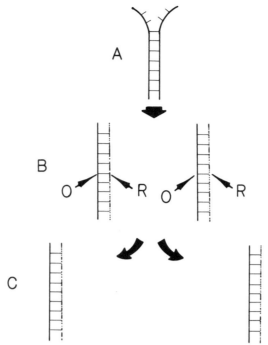

Figure 1.14. *The pattern of semi-conservative DNA replication. A. At the start of replication, the original DNA splits. B. Each original strand (O) is replicated (R). C. The new DNA combination of part old and part new DNA is distributed into the daughter cells.*

cell. The word "semiconservative" refers to the partial keeping of old DNA. At the end of the S phase, the cell enters the G2 phase with the nucleus ready for division.

In the G2 phase, the nuclear instructions must have already reached the cell cytoplasm, and any special proteins used in cell division may be made at this time. Still, cell biologists have up to now found no distinctive characteristics to impart to this phase of the cell cycle. On the other hand, it is quite reasonable to assume that some protein preparation would be going on during this period. So far, this is only an assumption.

During the G1 phase, the content of the nucleus is hard to see under the microscope. Even with special dyes and stains, the interior of the nucleus appears rather diffuse, with few indications of an internal organization other than the dark-staining nucleolus that appears as a tight, single spot within the nuclear field. But when cell division begins, all the special organization of the nucleus rapidly becomes visible.

First the DNA begins to condense and take on form under the microscope. Like dark rain clouds that condense out of the clear sky, these tightly wound molecules form dark-staining structures within the preparing nucleus. The structures are chromosomes.

When a technician removes the chromosomes from a crowded nucleus for a karyotype, the chromosomes display an unusual "X" appearance. The limbs of the X come from tightly coiled DNA molecules called *chromatids.* The point of joining the two chromatids is called a *centromere* and appears as a clear spot because it does not stain with the same dye used to color the rest of the chromosome. Figure 1–15 shows the relationship between chromatid and centromere for the chromosome.

The amount of DNA coiling that goes into forming the chromatids is considerable. Compacting DNA is like coiling and re-coiling an already coiled telephone cord. During this DNA aggregation, the DNA seems to be too wound up to make any messenger-RNA copies that could initiate other cell functions. All of the message-transcription needed to carry on cell replication and DNA replication must already be in

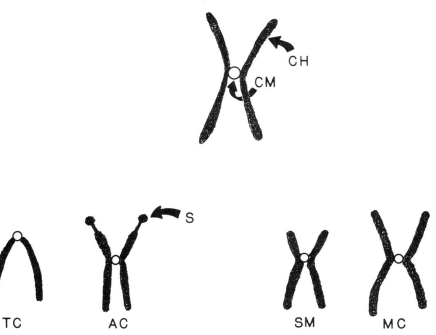

Figure 1.15. *Basic chromosome organization for replication. The chromosomes (CH) form out of the dissolved nuclear DNA at replication, connected together by a clear-staining structure called a centromere (CM). Different configurations include teleocentric (TC); acrocentric (AC); submetacentric (SM); and metacentric (MC), sometimes with satellite DNA (S).*

the cell cytoplasm before DNA is twisted into the chromatids to form chromosomes. Chromosome formation just precedes cell division, but the final division mechanisms must be set into motion.

The cell will physically separate nuclear material for the two new cells by mechanically pulling things apart, which sounds more haphazard than it really is. The pulling tools in the cell come from another bacteria-like inclusion that carries contracting segments for this process. This organelle is called the *centriole.*

The Mechanics of Separation

Under the microscope, centrioles look like cylinders with a hollow set of tubes arrayed in a circle forming the cylinder wall. The internal arrangement is nine pairs of tubes on the outside wall, and one pair down the center of the cylinder [31]. Like mitochondria, centrioles have their own DNA and can replicate as needed in the cell without using cellular DNA. And like mitochondria, they possess additional properties much like bacteria. The centrioles usually come in pairs that migrate to opposite sides of the cell when mitosis begins. At these command positions, they change size and shape and are renamed spindle formations. One end of the spindle formation attaches to the cell wall and the other extends out from the centriole body toward the chromosomes. The contracting limbs of the spindle are so fine that they appear to be no more than "lines of force" under a microscope (see Figure 1–16). The lines extend out from opposite ends of the cell toward the chromosomes. These lines of force are in reality contractile limbs that will physically pull the chromosomes apart.

While the centrioles are changing into their new form, the nuclear membrane dissolves away from the nuclear DNA. The chromosomes align along a line near the center of the cell. The limbs from the centrioles appear to attach to the centromere on each chromosome. The chromosomes then separate at the centromere, and the spindle-limb pulls each chromatid to opposite sides of the cell. The cell membrane pinches down to a fine thread between the two nearly formed cells and finally breaks. We now have two new cells, each with the nuclear material necessary to run its own life.

In each new daughter cell, DNA assembles near the cell's center, and the cell forms a new nuclear membrane around its DNA. The chromatids dissolve and disappear from view, and the cell nucleus is again clear, except for the dark spot of a nucleolus. While all this is going on, the mitochondria distribute between the two cells, often unevenly. They begin to replicate to increase the number of mitochondria in each new cell necessary to supply each cell's energy needs. The centrioles replicate themselves once, so each new cell has two, and the daughter cells finally slip quietly into the G1 phase and rest.

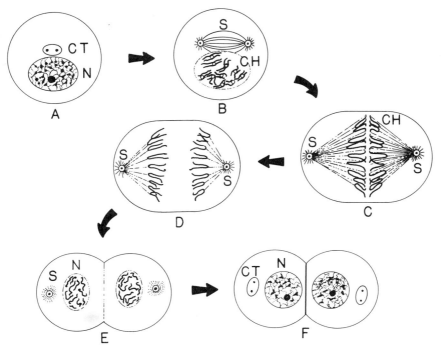

Figure 1.16. *The stages of cell division. A. The cell contains the two primary structures, the centrosome (CT) and the nucleus (N). B. The chromosomes (CH) form and the centrosome becomes the primary spindle formation (S). C. The spindles migrate to opposite sides of the cell and begin to mechanically pull apart the DNA. D. The spindles pull the DNA to opposite sides of the cell in preparation for cell splitting. E. The spindles decrease in size and two early nuclei (N) are formed with the cellular DNA, while a cell membrane forms to separate the two cytoplasms. F. The two daughter cells form normal nuclei (N) and two new centrosomes (CT).*

Just how long is this cycle for a cell? It varies according to the tissue. Let's look at some examples.

For the human fibroblast that makes connective tissue in the body, the cycle is about 53 hours long, with the following time partitioning: 40–42 hours in G1; 2–4 hours in S; 2–6 hours in G2; and 80 minutes in D [32]. Perhaps a better clue to the range of time distributions is the cycle times measured for various cells lining the mouse gastrointestinal (GI) tract. The cellular cycle-time parcels look like this: 17 hours for cells from the ileum (the end of the small intestine, just ahead of the large intestine), 36 hours for the colon (the large intestine), 85 hours for the lining of the mouth; and 181 hours for the esophagus (the tube between the mouth and stomach) [33].

Although the total cycle-times are quite variable, S, G2, and D for the mouse GI cells have the same length in each cell. The only variable is the length of the G1 phase. It is, therefore, the length of this phase, the resting phase, that controls the length of the cell cycle. We may not all agree on a single defining event to mark G1 in the cell's life, but something very subtle yet central to cell life occurs in this period.

The events in mitosis and DNA replication provide evidence that processes in the cell are governed by the same physical laws that determine physical events outside the cell. Cellular division requires: energy, provided by the mitochondria; a physical condensation of DNA into chromatids and chromosomes; and contracting centrioles that "muscle" the chromosomes apart and equally distribute the DNA into the two new cells.

REACHING FOR A FINAL FORM

At birth, we normally arrive in the world complete with all the organs and tissues required to keep our bodies working properly. Maturing toward adulthood causes some of these organs to change shape, but most of them simply become larger, while keeping the same basic form and function. Even in those tissues that do change with maturation, the cellular form is still present, and physical tissue qualities will still tell a histologist what the tissue is under a light microscope.

The more evident changes we see in the maturation process occur in tissues associated with primary and secondary sexual characteristics. Beards, body hair growth, a new body shape, and genital development mark the transformation. It is the last step in a cellular differentiation that started at conception in the uterus ten to thirteen years earlier. The body organs and tissues present at birth all come from a single cell with neither the shape nor the function associated with any final organ. The process by which the early cells of gestation transform from general, undedicated cells into the focused cells of an organ or tissue is called *differentiation.*

Differentiation refers to a functional specialization that often makes the cell dependent upon the rest of the organism to supply functions or capabilities given up to manage a dedicated task. For instance, the liver provides the brain with glucose from the liver's own glucose stores because the brain is unable to store its own glucose.

In the nucleus, differentiation means something else. The differentiation seems to shut down portions of the total DNA constitution, and large segments of the DNA are placed in storage [34]. The proteins that govern any cell's life then come from only those portions of the coding that pertain to functions that support the cell's specialization. This prevents, for example, contractile proteins that are normal to muscle from appearing in liver or nerve cells inappropriately.

Differentiation occurs in part through a complex process called *induction,* whereby a tissue proceeding down a well-defined path of specialization will "induce" cellular specialization in another tissue

[35]. Evidence indicates that the induction process is probably controlled through hormones, that is, molecules that have special message-meanings for both the releasing cell and the receiving cell. These hormones act only over short tissue distances and may exist for as little as a few brief seconds. These communicating molecules seem to turn off certain portions of the genetic code and prevent making messenger-RNA from selected DNA segments rather than turn something "on" [36]. The communicating molecule is called a "genetic repressor," for it represses gene expression in the receiving cell. Genetic suppression defines the code available for copying within a cell nucleus, and thereby, the tissue that develops from the cell.

The steps in differentiation that occur in gestation define each of us in a unique way. Our uniqueness comes first from the genetic code, and second from the cellular ability to express that code. Code-interpretation can be subject to several subtle influences. For example, chemicals from outside the cell or improperly formed molecules inside the cell can stop the full expression of a genetic code. An environmental contaminant can prevent normal spinal-groove closing in a developing fetus. The fetus then arrives with an open spinal column, which produces lower body paralysis and sometimes mental retardation. The condition is called *spina bifida*. We are, therefore, the product of two processes: first, the form of the genetic code (called the *genotype*); and second, the ability to correctly interpret that code into information the cell can use (called the *phenotype*).

KEEPING CELLS HEALTHY

To function, a cell needs energy, and that energy must come from "food." The word food has a specific meaning here, and refers to molecules that the cell can chemically dismantle through some energy cycle to gather chemical energy in the form of ATP. Earlier in this discussion, the mitochondrion appeared as a major transducer for this sort of energy conversion. Because the cell operates almost entirely on molecular ATP, were the mitochondrion to disappear and fermentation not be allowed, a cell would quickly die. The first requirement for cells to function is energy, chemically transduced into something the cell can utilize.

Events inside the cell are largely chemical, which means that things happen through the interactions of molecules and ions. The term "biochemistry" expresses those processes that are not simply chemistry, but a chemistry that occurs exclusively in biological systems. Like other chemical processes, biochemistry is temperature-sensitive. Reaction rates increase with an increasing temperature, and conversely, lowering temperatures can slow reactions. A familiar example is a slowing of muscle contraction when the muscle is too cold. A face exposed to a cold winter wind will quickly cool the underlying biochemistry of the muscle contraction. Until the muscles warm, smiling and talking require a great deal of effort.

Because the cell's chemistry is temperature-sensitive, the body holds a fairly regular temperature by making internal heat and controlling the rate of heat loss. Organisms that make internal heat are called *endotherms*. Animals such as snakes and lizards control their body temperature using heat from outside the body. They are *ectotherms*, and are commonly called "cold-blooded," although they really are not. Ectotherms, in fact, have body temperatures close to endotherms, but they control body temperature by using the environment, sun and shade. If the temperature outside the body should decrease, however, so does the body temperature of an ectotherm, and then its blood can indeed be cold. This temperature sensitivity observable in both the body and the test tube is an unmistakable clue that cells function under the same physical and chemical rules that are part of the world outside the living cell.

The cell needs not only the right internal environment, but also the right external environment. The environment of interest here is more than the world outside the body, but the solution of molecules and ions outside the cell, yet still inside the body. Essential elements such as sodium, potassium, chloride, molecules like water, and many other complex molecules contribute to cellular function in a variety of ways. Magnesium, we learned earlier, is central to ATP effectiveness. And the cell must also have a solution of sodium chloride to bathe its exterior.

Because none of these materials are made in the body, the organism must spend time and energy collecting the molecules it cannot make, and carry rather sensitive and sophisticated mechanisms to control the bathing solutions for a temperature and mix best suited to cell life.

Cells and most living things need oxygen to live. Oxygen, we learned, contributes to the mitochondrial synthesis of ATP. Deprived of oxygen for any length of time, cells quickly die, some faster than others. Nerves, for example, are very sensitive to the amount of oxygen present and they can disfunction from too much as well as too little oxygen.

In most organisms made of more than several million cells, the whole cellular community receives oxygen by transporting it in a fluid moving through a pattern of vessels or channels. Whether we are looking at a fly or a man, this fluid plays the same role of bringing essential oxygen molecules to the cells within the body.

CELLS OUT OF CONTROL

A recent Gallup poll indicated that the greatest current health fear in America is cancer. From our new vantage point for viewing the cell and its activities, it is easy to see the cancer cell as a cell out of control. The normal mechanisms that regulate growth are, for reasons yet to be understood, not functioning properly. The cancer cell seems to divide spontaneously, without regard to body needs or function. The rapid multiplication causes invasion and organizational destruction of organs, leading to illness and death in the absence of some sort of intervention such as surgery, chemicals, or radiation.

About 100 clinically distinct types of cancer exist [37]. Each has a form of therapy that makes it distinctive and recognizable from the others. All these cancers can be classed under four major headings: first, *leukemias* that produce abnormal numbers of white blood cells in the bone marrow; second, *lymphomas* that produce abnormal numbers of white blood cells in the spleen and lymph nodes; third, *sarcomas* that are solid tumors stemming from connective tissue, cartilage, bone, muscle, or fat; and fourth, *carcinomas* that are solid tumors that derive from surface or epithelial tissues (skin and body cavity linings). A primary feature of normal cell division noted earlier is the phenomenon of contact inhibition. Cell contact inhibition puts crowded cells that have a lot of membrane contact into the G1 phase of the cell cycle for long periods of time. Cancer cells have lost this inhibition. The results of a couple of experiments will show how these inhibitions normally work and fail in cancer.

Normal cells, removed from the body and retaining their ability to multiply, will form a growing population of cells if they sit in a special solution with nutrients and oxygen at the correct temperature. This laboratory growth of cells is called *cell culturing* or sometimes *cell plating*. When the cells fill the culture dish, the population stops growing because of contact inhibition.

Repeating the same experiment with cancer cells produces results that are almost but not quite the same. The cancer cells also fill the culture dish, but the population does not stop growing. The cells do not stay in the G1 phase, but immediately go on into the S phase and the remainder of the cell cycle. Contact inhibition, which normally limits the growth of tissue in the body, is somehow disabled, and just as they crowd one another in a culture dish, malignant cells invade and crowd out normal tissue in organs carrying the malignancy.

This continuous growth in the body causes cancer cells to take in energy in large amounts [38]. Under normal conditions, cell metabolism is the source of heat used to maintain body temperature. Cancer cells, by consuming large amounts of energy, give off even more heat.

This higher-than-normal cellular temperature permits detection of some solid tumors by looking for hot spots on the surface of the body, using very sensitive heat-detecting instruments [39]. But because of the prodigious amounts of food required (the food is diverted into energy and molecules to make more cancer cells), another process begins to take over the cancer victim faster than cellular invasion.

Cancer cells successfully divert body-energy intake and the body energy-stores for their own use, even diverting the normal vasculature to support growth [40]. As a result, the normal body tissues suffer malnutrition. The effect is called *cachexia,* and the result is general body-wasting and ill health. Body-wasting and all the complications of malnutrition stress the normal tissues severely. Cancer patients more frequently die from the complications of cachexia than from the invasive effects of the cancer [41].

Cancer cells in culture not only lack contact inhibition, but also never stay very long in the G1 phase. The cellular controls for multiplication and growth appear as instructions for the cell during this period, and this is also the time in which RNA carries the DNA messages for the cell from the nucleus to the working machinery in the cell's cytoplasm.

The loss of control in the G1 phase can come from several potential sources. One way a cell can lose control is through a viral infection that does not destroy the cell. In this nondestructive infection, the viral DNA infects the cellular DNA, transforming the cell into a potentially abnormal cell. Because the viral DNA is incorporated into the cellular DNA, it is replicated right along with the rest of the cellular DNA with each cell division. Thus, each cell division spreads the infection to each of the daughter cells. The final step that might make this cell cancerous is, unfortunately, unknown. And although this process can be connected to several animal cancers, it seems not to be a major source of cancer in humans [42]. Only one cancer, Burkett's lymphoma, is directly connected with a viral infection, and only a few others suggest some sort of viral involvement.

The majority of human cancer-producing mechanisms are chemical, and nearly all come from contaminants in the environment we live in [43]. Chemicals that cause cancer, called *carcinogens,* are found in many industrial wastes. They include substances such as formaldehyde and nitrites. Carcinogens also appear in nonindustrial pollutants like cigarette smoke.

The remaining source of cancers is ionizing radiation that alters the cell nucleus in an unpredictable manner. One example of such radiation is ultraviolet light from the sun. Ultraviolet light interacts directly with cellular DNA. Because the skin is on the outside of the body, the most frequent target is the DNA in germinal skin cells. These germinal cells divide to make the layers of dead cells that give us a water-tight covering on the body. If the ultraviolet light causes the appropriate alteration of the DNA in these germinal cells, the result can be skin cancer. Other sources of ionizing radiation include radioactivity that comes from natural substances in the ground, industrial accidents involving radioactive materials, and the medical use of X ray [38]. All can produce cancer, often in deep body structures.

But cancer may not be as "unnatural" as we might like to think. Recent findings in cell physiology suggest an engaging hypothesis about the generation of cancer cells as part of normal body function [40]. The hypothesis suggests that in any population of cells, the sort of DNA damage that could produce a loss of cell control has a finite probability of occurring. When we consider the astronomical number of cell divisions that are normally part of our daily existence, turning up a cancer cell may be much easier than we imagine. For example, nearly 2.5 million red blood cells must be made every second in the body to keep the red cell population close to normal. The number of cell divisions in a body each day is beyond convenient expression or thinking. The possibility of one cell being made incorrectly, and therefore becoming cancerous, is really quite good. This comes about not because the probability of producing a cancer cell is high, but because so many cell divisions occur that even very small chances include a physically large number of cells. These altered cells, however, appear abnormal to the body, which detects and methodically destroys them using the body's immune response.

This hypothesis gathered support because of evidence that spontaneous remissions (in which the disease begins to decrease and ultimately disappears for no known reason) occurred in some cancer patients who, for other reasons, had their immune systems stimulated [44]. Cancer is also a significant cause of death in kidney transplant patients whose immune systems have been chemically suppressed in order to prevent their bodies from rejecting the transplanted organ [45]. A detailed understanding of the process is not in hand and much work remains to be done. Nevertheless, these clues to body functions may offer better ways of dealing with cancer.

CELL RESPONSE TO INJURY

Just as the cell is the center of life, it is the center for disease. Events that produce disease cause cellular and subcellular alterations that can be seen with the microscope and in some cases, detected biochemically. It is the cell's response to these alterations that leads to the conditions we call disease [46].

A healthy, normal cell has a wide range of adaptation, within which perturbations can produce injury to the cell without producing cell death. The forces that cause the perturbations could be lethal or sublethal for the cell, and yet produce cell injury without cell death. This is true so long as the perturbations do not force the cell beyond its range of adaptability. The cell is like a tightrope walker, permitting alterations that do not lead too far from equilibrium. In general terms, then, an injury is anything that alters or upsets the normal balance of events within the cell [47].

The sources of injury are several. They can come from inside a cell, through either a bad DNA template or a faulty transcription from a good template. Injuries can come from the outside in the form of malnutrition, chemical poisoning, microbiological agents, physical agents, immune responses, or oxygen deprivation. All of these processes produce their effects by moving the cell away from its equilibrium condition.

More common forms of cell injury and cell death come from bacterial or viral infections. Bacterial infections involve toxins that alter or destroy normal cell functions. One of the more sensitive portions of the cell is the cell membrane, which can be altered in both structure and function by bacterial toxins. Viral infections, on the other hand, usually involve destruction of the cell as a virus takes over the cellular machinery to create more viruses, ultimately bursting and killing the cell.

Another common form of cell injury is oxygen deprivation. Lowering the amount of available oxygen for a cell is called *hypoxia*. When no oxygen is available, the condition is called *anoxia*. Because oxygen is supplied by the vascular system, a reduction of blood supply, either partial or complete, can produce a condition of hypoxia, leading ultimately to anoxia. A reduction in blood supply is called *ischemia*. A heart attack is an example of an ischemic condition in the heart's vascular system from either a partial or a complete reduction in blood flow. The initial ischemic condition produces hypoxia; and if the reduction is complete, it produces anoxia and cell death.

All cells do not respond to hypoxia or anoxia the same way. For example, a neuron deprived of oxygen will die in 2 to 4 minutes [48]. A heart cell, on the other hand, can withstand oxygen deprivation for a period of 15 to 20 minutes [48]. In contrast, portions of the kidney can withstand anoxia for 1 to 2 hours at body temperature before cell death occurs [48]. And all of these times can be extended by reducing the cell temperature and thereby its requirements for oxygen. This temperature-dependent response to hypoxia explains how cold-water drowning victims can survive for long periods of time without oxygen with little or no damage to the brain, and permits the use of hypothermia (lowered body temperature) for certain operations.

Central to the concept of the cell as a source of disease is that all cell responses as part of disease are already within the cell. Disease introduces no new elements in cell biology, but simply accentuates or removes normal cell processes. It is in this sense that we see the cell as the source of both life and disease.

CONCLUSIONS

We are a mosaic of cells, each dedicated to a function that contributes to our existence. Some of these cells will be examined in some detail in the following chapters as we look at the controls and communications of physiology. And all is possible because cells are alive, and in their life and organization, they bring us life. The internal mechanisms of the cell are molecular, so an organism must supply the molecules needed for cellular energy and any molecules it cannot make. Malnutrition results should it fail to do so.

Still, the cell is quite adaptive and will shift its own biochemistry to keep things going as long as possible—in other words, to keep its biological equilibrium. If proper molecules do not come from the outside, the body will draw upon its own supplies and sacrifice certain body tissues for survival of the whole. When events push the cell too far, it may be injured or even die, and its response produces what we call disease.

Guarding the cell's interior is a very thin yet complex membrane. Long thought to be just a confining structure, it is instead a very active component of the cell. And it may be that through this flexible membrane will come an understanding of cancer. When we look at cell components, it seems that even the simplest of structures are active contributors to the complexity of cell life. The cell indulges no freeloaders, except for some types of viral infections.

When cells multiply in a controlled fashion, they bring us function, structure, texture, and life. When they lose control, they can bring pain, suffering, and death. Yet these small agents of life walk a defiant path in this universe. While the rest of the universe is running down and becoming more random, the cell self-organizes, remaining highly structured. In simple logic, the cell's own chemical reactions should breed randomness and death within. Instead we see life all around us, deep-seated, unyielding, cellular.

2

MUSCLE AND MOVEMENT

MUSCLE IN THE LARGE

Unlike the single celled animal called an amoeba that moves by extending a portion of its cell and flowing into it, most complex organisms, including humans, move with the aid of specialized contracting cells called the skeletal muscle cells. These cells are collected into aggregates called "working units," further collected into larger groups called muscles, and strategically draped over a bony skeleton. Muscles permit joints to bend and extend, limbs to rotate, and fingers to move in precise arcs. To achieve the myriad of motions available to the body, a human needs a lot of different muscles because they work only in contraction.

Because a muscle can only contract, a specific muscle or group of muscles must exist for each basic movement, and that sets muscles working in opposition to one another. Biologists call this opposition an antagonistic relationship. The muscle doing the contracting is called an *agonist*. The muscle resisting the agonist's contraction is called an *antagonist*. Through controlled contraction of the agonist and controlled release of the antagonist, limb motion is smooth and steady with few visible jerks.

The qualities of human motion and a muscle-modulated body form were objects of appreciation long before the early researchers revealed even the most basic mechanisms of muscle contraction. Artists' attraction to the human form has a tradition that reaches from the time before the Greeks to the abstractionists of today. Although da Vinci dissected the dead to understand muscular anatomy and painted well-muscled bodies, he had not the faintest notion of how muscle worked. In their ignorance, wise men of his day attributed contraction first to the tendons, reasoning that the muscle just went along for the ride; then they improved their knowledge by attributing contraction to fluid movements and mysterious "humors." The advent of electronmicroscopy, sophisticated biochemistry, and X-ray crystallography brought out the true elegance of muscle contraction. Even with these powerful tools, mysteries still abound. As with drilling for oil, all the easy finds have been made. The remainder await new technologies and new techniques for disclosure.

Skeletal muscles are normally under neural control from the brain and spinal cord. Control for skeletal muscles comes from sensing what they are doing and returning that information to controlling areas in the brain. Sensors within muscle units tell the brain what is happening to a muscle by sending information on its tension and rate of tension development back to the brain. Additional information comes from other sensors in the joints that relay joint position and bending to the brain. It is a good system, offering the human form a wide range of muscular activities without injuring or tearing muscles in use.

But we are still no closer in our discussion to the mechanisms of contraction than da Vinci was. Before jumping into the contractile process, let's take a general look at the tissues called muscle.

THREE KINDS OF MUSCLE

Although "a rose is a rose is a rose," the same cannot be said of muscle. Muscle is a general term used to denote those tissues in the body that contract as a specialty. The division of muscle could become quite detailed, but general observations easily divide muscle into three kinds: smooth, cardiac, and skeletal. This division is more than theoretical, for histologists use three basic criteria for sorting out types of muscle. These criteria are: first, the muscle's location in the body; second, its appearance under the microscope; and third, the effect of a muscle contraction. Looking at these criteria, it seems that our division is based largely on anatomical considerations. This is true, and for good reason. Anatomy was the first real biological science, and its tradition remains firmly entrenched in today's biology.

A search for smooth muscle shows that it appears in motions that are both rhythmic and automatic. For example, the walls of arteries, intestines, and both the gallbladder and urinary bladder are well supplied with smooth muscle. Nerves that innervate smooth muscle

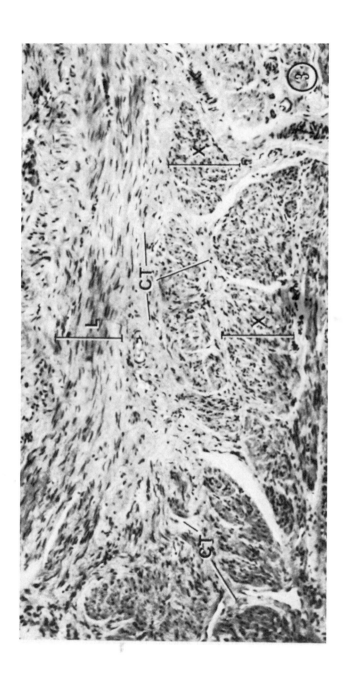

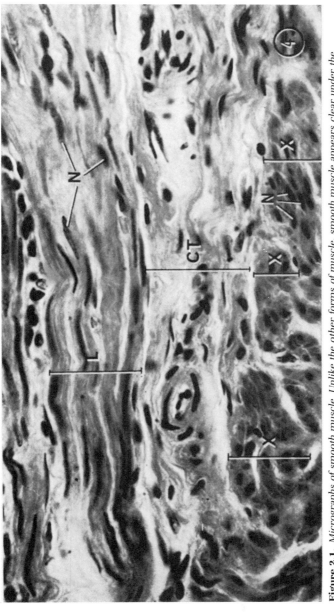

Figure 2.1. *Micrographs of smooth muscle. Unlike the other forms of muscle, smooth muscle appears clear under the microscope. N are the smooth muscle nuclei; L is a longitudinal cut through the muscle tissue; CT is a connective tissue surrounding the smooth muscle; X is the smooth muscle bands in cross section. These two sections are from the human uterus. By permission, Edward J. Reith and Michael H. Ross, Atlas of Descriptive Histology, Hoeber Medical Division, Harper and Row, New York, 1965.*

come from the autonomic nervous system. Despite this generous supply of innervation, smooth muscle has an ability to operate without nerves, and will contract in a rhythmic fashion in the absence of any nerves.

Smooth muscle gains its name from its internal cellular appearance under the microscope—it is smooth and clear, without the many crossbands or striations of the other muscle tissues (Figure 2–1).

Cardiac muscle, as the name connotes, is found in the heart and seems to share characteristics of both skeletal and smooth muscle. Cardiac muscle contracts ultimately to pump blood throughout the circulatory system. Like smooth muscle, the cardiac cell can contract in a rhythmic fashion with or without nerves. Like smooth muscle, the cardiac cell is small and has a single central nucleus. On the other hand, cardiac cells have striations and an internal organization much like skeletal muscle (Figure 2–2). The cardiac muscle cell seems to have taken the best qualities of both the other two muscle types and combined them to form a new muscle cell.

Skeletal (or striated) muscle, the real object of this discussion, obtains its names from: first, its location (attached to the skeleton); and second, how it looks under the microscope (striated). The cellular crossbands that are called striations come from a specific internal, molecular organization (Figure 2–3).

In the physical world, crystals made from the repeated organization of molecules are some of the most highly organized structures in our experience, with a degree of organization not expected in the biological world. Nevertheless, the visible striations found in skeletal and cardiac muscle come from a molecular organization that can rival the natural crystals for regularity. Like a natural crystal whose shape mirrors its molecular organization, the internal organization of skeletal muscle reflects the organizational properties of the whole muscle. Unlike rigid crystals, however, the muscle cell keeps its regularity and a delicate flexibility along with a remarkable ability to contract.

All three types of muscle tissue normally receive control from nerves. The autonomic nervous system controls smooth muscle and cardiac muscle in a largely unconscious manner—we do not have to think about these functions to have them happen. Yet these two muscle tissues can continue to work and adapt to changing demands without neural influences. In contrast, controls for skeletal muscle come from another part of the nervous system that does demand conscious activity. Interruptions in this neuro-control disrupt muscle function. Without nervous activity, skeletal muscles become limp and useless.

The nerves extending from the brain to the skeletal muscles are called motor nerves, and as we shall see, these nerves play a unique communicative role.

Figure 2.2. Low-power electron micrograph of cardiac muscle. The contractile proteins are clearly visible, with the dark insertions of the proteins into the ends of the cardiac cells. Mitrochondria are distributed between the filaments and along the left edge of the micrograph. By permission, William Bloom and Don W. Fawcett, A Text Book of Histology, Tenth Edition, W.B. Saunders Co., 1975.

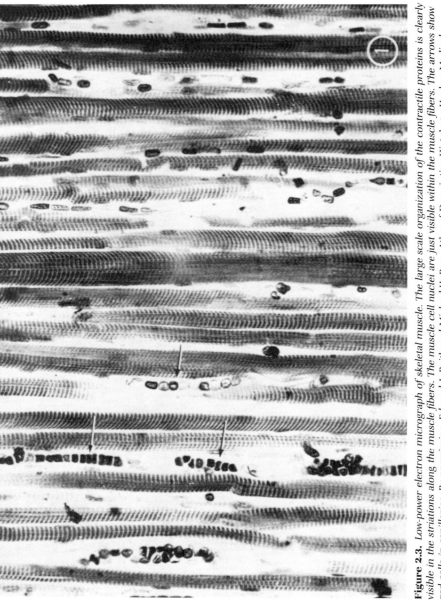

Figure 2.3. Low-power electron micrograph of skeletal muscle. The large scale organization of the contractile proteins is clearly visible in the striations along the muscle fibers. The muscle cell nuclei are just visible within the muscle fibers. The arrows show red cells in capillaries. By permission, Edward J. Reith and Michael H. Ross, Atlas of Descriptive Histology. Hoeber Medical Division, Harper and Row, New York, 1965.

It is time to look inside the muscle cell to see how the neural messages are turned into physical contractions.

SKELETAL MUSCLE CELL ORGANIZATION

Skeletal muscle is made of long thin cells, some of which span the complete length of the muscle. As cells go, this is a large cell, and it has some unusual ways of handling this larger size. A cell that extends over several centimeters with a proportionately large diameter will need more than one nucleus to control a unit volume of the cell if the rules found in other cells still apply. And here the skeletal muscle cell follows the rules. Regularly spaced along the muscle cell, hugging the cell wall, are nuclei, each controlling a cell volume very close to the same volume found in other body cells [1]. To be sure, a skeletal muscle cell looks unique, but it follows the same physiology as many other cells in the body.

With so many nuclei in a single cell, the question quickly arose: Did the muscle cell come from one cell or from many cells fused together? Some carefully conducted experiments gave the answer: The skeletal muscle cell comes from the fusion of many single cells, each with a single nucleus, forming what is called a *syncytium*, a single cell with many nuclei [2].

Along with the many nuclei, the cell also contains mitochondria that supply energy, a strange-looking endoplasmic reticulum, and large crystal-like proteins that make up the contractile machinery. It is these proteins that let the cell contract.

THE CONTRACTILE MACHINERY

A single muscle cell is easy to dissect away from its companions and suspend in distilled water. Because of the difference in osmotic pressure inside and outside the cell, the distilled water flows through the membrane into the cell. Soon the cell membrane bursts under the relentless growing pressure within. When the cell membrane bursts, the contents spill out into the water suspension, and the cellular contents are open to inspection under a microscope.

Among the many cell components that spill out are long threads of contractile proteins. These threads are called *myofibrils*. The striations that give the tissue its name are regularly distributed along the length of the myofibril. At this point, vision is at the limit of resolution for the light microscope. The electron microscope provides further revelations about the organization of the myofibril.

The Biochemistry
of Contractile Proteins

Before the electron microscope probed into the anatomy of the myofi-bril, biochemists had already found two proteins in large quantities within skeletal muscle. Investigators named them *actin* and *myosin*. While in solution, some important properties of these proteins came to light. The ever-probing microanatomists and the relentless biochemists discovered and fitted together the pieces of the organization like a three-dimensional jigsaw puzzle. Here are the results as we presently understand them.

In the beginning, there was confusion. Two entirely different kinds of actin appeared in the cell. One was globular in shape and was called G-actin. The other was in the form of a long thread or filament called F-actin. It turned out, however, that G-actin was the precursor of F-actin, and stringing the G-actins together produced the F-actin [3]. This connection appears in the following chemical equation:

G-actin + ATP = F-actin

ATP stands for adenosine triphosphate, and in a broad sense, it also stands for energy. Order emerges from confusion.

The Stereochemistry
of Contractile Proteins

Functional actin, in its native form, has two F-actin strands wound around one another to form a slow-twisting double helix. Because these two strands act as a unit, it is easy to consider the helix to be a single entity. The actin strands are further organized into three-dimensional arrays, like needles extending out from both sides of a flat surface. A two-dimensional schematic representation of the actin organization appears in Figure 2–4. The filaments connect at the Z-line in a nonsymmetrical way. This kind of organization is called a *protein array* and represents the cellular actin organization.

The second protein of interest is called *myosin*. Myosin is a very large protein molecule composed of many smaller protein subunits. The myosin protein looks like a log with a lot of short, blunt limbs sticking out. A schematic is shown in Figure 2–5.

The limbs are made of another protein called *meromyosin*, which has a unique task.

Myosin, like the actin, is held in an array without an apparent surface of attachment. The centers of the molecules are in common plane, and some electronmicrographs suggest some sort of attachment at the centers via a very fine molecular net.

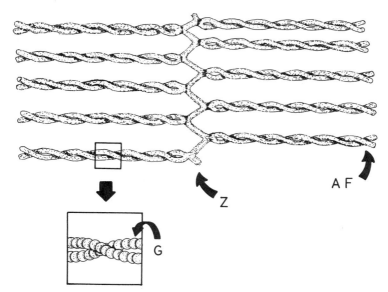

Figure 2.4. *The primary arrangement of actin filaments. The globular form of actin (G-actin) (G) are polymerized into filaments of F-actin (AF). The filaments extend outward from the Z-line (Z), and are formed by twisting two G-actin filaments together.*

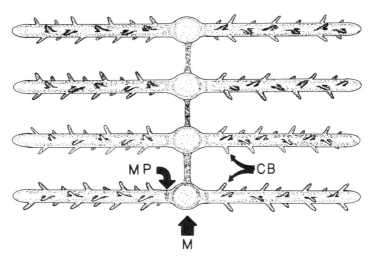

Figure 2.5. *Organization of the myosin protein. The myosin (MP) filaments are made of a smaller protein called meromyosin, which forms the cross bridges (CB). The myosin array is joined together with a thin filament down the center, forming the M-line (M). Myosin joins with actin to form the complete contractile array.*

The myosin and actin array fit together in such a fashion that they can slide past one another. The intersection is much like two brushes with their bristles pushed together. The myosin array is effectively double-sided, and an actin array appears on both sides of the myosin. The process of intersection is called *interdigitation*, and the final organization is shown in Figure 2–6.

The Sarcomere in Action

The final molecular organization is the basic unit of contraction called the *sarcomere*. The sarcomere is defined to be that portion of the myofibril extending from Z-line to Z-line. These sarcomere boundaries are shown in Figure 2–6.

Two additional regions of the sarcomere take on importance to this discussion. They are the region called the A-band, which spans the length of the myosin molecules, and the I-band, which extends from the edge of one myosin array to the next. They are also shown in Figure 2–6.

While observing a myofibril under a microscope and stimulating contraction, an observer would see the following events: first, the Z-lines come closer together; second, the I-bands become smaller; third, the A-bands remain unchanged. Under the electron microscope, the "bumps" on the meromyosin appear to attach to the actin threads to

Figure 2.6. *Formation of the sarcomere. The functional unit of contraction is the sarcomere (S), formed by the interdigitation of actin filaments (A) and myosin (M). The sarcomere (S) extends from Z-line to Z-line (Z). The array forms the A-band (AB), which spans the width of the myosin array, and the I-band (IB), which spans the distance between succeeding myosin arrays.*

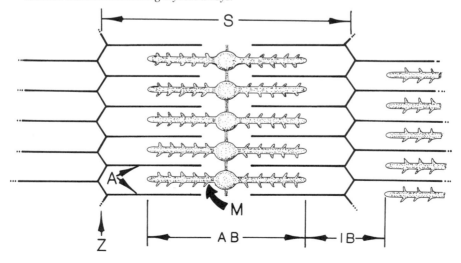

form what are called cross bridges. Based on these data, let's simply consider how the process of contraction might occur.

The proteins could shrink in some manner, but the A-bands remain unchanged during contraction, so the meromyosin is not shrinking. On the other hand, the actin might be shrinking because portions of it extend into the A-band, and shortening the I-band could also mean shrinking the actin. But detailed electronmicrographs indicate that actin is not shrinking to any detectable degree. Still, the proteins could fold in some way, but again, the microscopic and biochemical data exclude this possibility. The only model that follows from the evidence and satisfies all the observations is one in which the proteins slide past one another like interdigitating fingers [4]. This contraction model is called the sliding filament model and successfully predicts many of the observed properties of muscle, including both cardiac and skeletal muscle.

Before continuing further, the use of the word "model" deserves some explanation.

Physical events in both living and nonliving systems are often very complex. Scientists studying an event like to reduce the considerations to only the essential and develop a fundamental concept of what is going on in a simplified form. The result is called a model. A model, then, is a simplified description representing the essence of a physical event or process. A model also represents a scientific hypothesis that is subject to the same continual verification and improvement as any hypothesis. If the model can successfully predict results from experiments on the real system, the model becomes a way to think about events and design further experiments to find out more about what is happening. Our sliding filament model, therefore, is an essential yet accurate description of muscle contraction.

Developing Tension

Because the physics of a model must be consistent with the basic rules of physics, physical events in the model spring from relationships often understood from larger systems. For example, for the sarcomere proteins to slide past one another, some sort of tension must form between the actin and the myosin. The only place of common attachment that could transmit such tension is the meromyosin cross bridge. The importance of the bumps that we took note of earlier now becomes apparent. The cross bridges must be the *source* of muscle tension. The exact mechanism is unknown, but no doubt exists that the meromyosin cross bridges do develop the tension needed to pull the Z-lines toward one another [5]. As the two protein arrays move, the cross bridges "walk" on the actin, each cross bridge making several attachments during a contraction. Protein is like a rope: it can pull objects (tension), but it cannot push objects (compression). The ropelike qualities of proteins begin to explain an earlier observation that muscles can work only in tension.

The functional distance from muscle to sarcomere is quite large. The sarcomere is only one very small unit of the myofibril, and hundreds of these myofibrils are in a muscle cell, and hundreds of muscle cells in a muscle. So, it appears that a very large number of chemical reactions occur for even the smallest muscle twitch. At the same time, the sarcomere is not very large. It may be as much as 1/100,000th of a total muscle length or less, which is hardly large enough in itself to bring about a measurable muscle contraction if a single sarcomere shortened. But muscle is not just one sarcomere; it is many sarcomeres acting in concert, lined up along the myofibril. When more than one sarcomere contracts, each will add to the total decrease in length. If all the sarcomeres contract maximally, the muscle cell can contract to its smallest length. At the same time, because the sarcomeres are working together, they reflect their properties to the whole muscle, just as the molecular organization of a crystalline material specifies the shape of the crystal.

So far, we have looked at only the all-or-none qualities of contraction. They do not explain the partial contractions that make up our own muscular experience. On the other hand, the tension developed by a cross bridge is probably an all-or-none event. By recruiting a various number of cross bridges and sarcomeres, the amount of tension and shortening developed by a muscle can be quite variable. The cellular rules governing the recruitment of cross bridges and sarcomeres are as yet unclear, but a fine modulation of muscle tension is possible, giving us a great deal of adaptability in movement.

Driving all these contractions and chemical reactions is energy. Before looking closer at the sarcomere, let's stop and look at the source of energy for contraction.

ENERGY FOR CONTRACTION

Nearly all biological events require some sort of energy to work, and this is true of muscle contraction as well. Some of these cellular events and the sources of energy appeared in Chapter 1. One more cellular event can be added to the list: cross bridge tension. Cross bridge tension can develop only if energy is available to do so. And it is adenosine triphosphate or ATP that brings energy to the cross bridge.

The molecule, ATP, is not of itself energy, but a transport system that delivers energy to many cellular events, including cross bridge tension. ATP is a little like the red blood cells that are not oxygen, but simply supply the means of getting oxygen to the tissues. In the red cell, oxygen is carried in the protein hemoglobin. In ATP, the energy is carried in the formation of phosphate bonds, stored as chemical energy. Let's take a closer look at ATP.

The ATP molecule is composed of four basic components, a large molecule called *adenine* (hence the first part of the name adenosine) and three phosphate groups. The adenine molecule seems to be both a base for the three phosphate groups as well as a source of other more subtle chemical influences. Along with the chemical characteristics, the whole molecule must be bent in a specific way for the energy to be released. The phosphate limb and the adenine base bend toward one another when another molecule, magnesium, becomes part of the complex, providing the string for this molecular bow [6]. The stored energy is available when the last phosphate group breaks off, and the molecule becomes adenosine diphosphate or ADP. Thus, for each bundle of energy released by an ATP molecule, the ATP separates into ADP and a free phosphate molecule. Now another control and transfer mechanism enters the scene.

An ATP molecule is inherently unstable and subject to spontaneous breakdown. The spontaneous release of ATP-bound energy is like a fire suddenly appearing in a fireplace on a hot summer day. The fire is good but not for a summer day; the fire needs to be conserved for a cold day. Accordingly, energy in a cell must be controlled if the cell is to use the energy effectively. The controller is an enzyme.

An *enzyme* is really a biological catalyst that speeds a chemical reaction in one direction or another. Unlike the small-sized molecules that make up the catalysts of inorganic chemistry, enzymes are usually very large, complex proteins with very specific shape requirements that enable them to work. This molecular specificity permits many different reactions to occur within a cell with little chance of one reaction cross-contaminating another. For ATP, the enzyme is called ATPase [7].

ATP is made within a muscle cell at two sites. One site is in the cell fluid outside the nucleus (the cytoplasm) and involves the metabolism of glucose in a manner very similar to the fermentation of glucose into alcohol by yeast [8]. The process requires no oxygen, and is therefore called *anaerobic*, and produces two by-products, pyruvate and lactic acid.

The second synthesis site is located in the mitochondria. The mitochondria consume oxygen to churn out ATP molecules that give the muscle cell energy to run on. This process uses oxygen and is, therefore, *aerobic*. The two sites of ATP production work in a complementary way. When energy requirements exceed the aerobic capacity, the anaerobic process begins to supplement ATP production [9].

Cross bridges need ATP-bound energy to generate tension, regardless of where it is made. To release the energy in a controlled fashion, each cross bridge contains an ATPase. Although the presence of ATP and an ATPase will not be enough to set the final mechanisms of contraction into motion, ATP and the cross bridges are pivotal.

ATP AND CROSS BRIDGES

When early investigators began to look at the relationship between ATP and cross bridge formation, what should have been an easy mechanism to grasp was just the opposite. The data immediately caused a controversy as different laboratories obtained conflicting findings. More recent work has cleared away some of the haze and a clearer picture of dependencies has emerged. The strange case of ATP, cross bridge bonding, and the development of tension must be considered from the unusual vantage point of cellular ATP concentrations [10].

If intracellular levels of ATP are normal, the sarcomere and indeed the whole muscle can contract on command to full potential. When muscle activity increases, the intracellular ATP quantities drop, and a muscle becomes fatigued; that is, its maximum performance level goes right down along with the decreasing ATP quantities. With no rest, a muscle will eventually reach a point of being unable to contract at all. With rest, the cellular ATP levels are restored, and with ATP, muscle strength returns [11].

In contrast to moderate drops in ATP quantities, ATP levels that drop too low produce an unexpected sequence of events. When ATP content becomes too low, increasing numbers of cross bridges begin to bind to actin in very rigid, inflexible bonds [12]. As ATP levels decrease further and further, more and more of these bonds form. This sort of bond is called a "rigor bond" because it is the bond found in muscle that enters a state of *rigor mortis*, the "stiffness of death." If ample amounts of ATP return to the cell before it dies, the rigor bonds can be reversed, and the contractile proteins once again become flexible tension machines. Thus, it appears that overall stiffness of a muscle depends markedly on the amount of ATP present in the cell.

ATP normally does not appear in muscle without its cohort, magnesium. Magnesium is a small ion with two electrons missing from its outer electron shell, which gives it a net positive charge. Because two electrons are missing, magnesium is called a divalent cation, that is, a positively charged atom with two electrons missing. Magnesium, because of its size and shape, binds to ATP in such a manner that it bends the ATP molecule to form a specific ATP-magnesium shape.

Also found in the body is a second molecule very similar to magnesium (but not exactly like it) called calcium. Calcium can also fit into the ATP molecule and bend it to a new shape. Notwithstanding the similar bend, something very subtle is lost in this calcium-ATP

60   MUSCLE AND MOVEMENT

combination, for it will not permit an energy transfer from ATP to the cross bridge [13]. The calcium-ATP complex is like a key that fits in a lock that cannot be turned. Only the magnesium-ATP combination will permit the energy stored in the ATP phosphate bonds to transfer to the cross bridges to become muscle contraction.

With ample amounts of ATP available to the cell, the cell can perform to full potential. ATP does not turn on contraction, however; it just provides the energy. ATP, therefore, is not the controller for the contractile process. We have to look elsewhere for this controller.

CONTROL OF CONTRACTION

An early observation about muscle contraction was that free calcium ions had to be present *inside* the cell for any contractions to occur. In contrast to the failure of calcium to function with ATP, calcium in its free ionic form must be present for the contractile machinery to shorten [14]. With ample ATP and no calcium present, the machinery remains unactivated, poised for a contraction that cannot occur. When the intracellular free calcium levels increase, however, contraction occurs. With subsequent removal of free calcium, the contractile proteins relax. Here, then, is an agent that controls contraction.

If calcium is the agent that tells the contractile machinery to work, where is the calcium stored when not in use? It cannot be allowed to just float around in the cell. Should calcium be always available, the cell would undergo frequent and continuously uncontrolled contractions. As in the case of ATP, uncontrolled energy release (contraction in this case) is counter to efficient cell function. The search for a calcium storage site directs our attention to another part of the cell, the sarcoplasmic reticulum.

The *sarcoplasmic reticulum* (SR) is a specialized version of the endoplasmic reticulum (ER) found in other cells [15]. In Chapter 1, it was clear that cellular ER normally provides a special synthesis site for cell components and secretions. In the muscle cell, its role is the storage and release of calcium. In the resting state, the SR vigorously pumps the calcium ions from the muscle cell's interior into storage sacs called *terminal cisternae* that are part of the SR.

On the cellular command to contract, the SR rapidly releases the calcium, turning on the contractile machinery. After the command passes, the SR begins to remove the free calcium from the cell's interior, causing the cross bridges to release and the contractile machinery to relax. But here an unusual puzzle appears. Some elegant experiments using a marine animal's phosphorous dye that gives off light when mixed with ionic calcium show that the calcium released from the SR and reaching the protein arrays is in an ionic state, that is, the calcium is charged and free to move. In contrast to the released calcium, the calcium *returning* to the SR and pumped into the SR is *not* a free ion [14]. What form is it in then? The answer is still unknown. Despite the mystery about part of the steps involved, calcium is clearly a principal controlling agent. The question now is: When calcium gets to the protein arrays, what does it do there?

CALCIUM-SENSITIVE MOLECULAR CONTROL

It was a difficult job to find out just what was happening in the contractile array each time calcium arrived on the scene. The discovery of two new proteins hidden in the contractile machinery paved the way to understanding. These proteins were named tropomyosin and troponin.

Tropomyosin is a long thin protein that normally hides in the groove between the twisting strands of F-actin [16]. This "fit" is illustrated in Figure 2–7.

Because of its small size and the small cellular quantities involved, tropomyosin carries on its important task in almost impenetrable secrecy. Tropomyosin sits on the F-actin threads, and by strategic positioning, blocks the bonding between actin and the meromyosin cross bridge. A bond can form only when tropomyosin is moved or removed. But this is not entirely true. Despite these control procedures, rigor bonds can form if the cellular ATP level is too low. When isolated, tropomyosin reveals a curious property: it does not bind calcium ions, making it only part of the control mechanism. The calcium binder, it turns out, is *troponin*, the second part of a two-stage trigger that effects contraction. Troponin, a protein smaller than tropomyosin and present in even smaller quantities, carries the sought-for calcium receptor sites. Troponin sits right on the tropomyosin molecule and in the presence of calcium, begins to influence the activities of tropomyosin

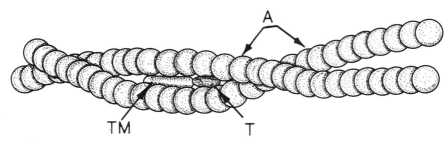

Figure 2.7. *Control proteins within the actin filaments. Two control proteins within the actin filaments (A), tropomyosin (TM) and troponin (T) fit within the groove formed by the two actin filaments.*

MUSCLE AND MOVEMENT

[16]. These two small, hard-to-find molecules provide an organizational relationship of unusual character.

In the resting state, no calcium is present at the contractile machinery, and tropomyosin sits over the actin-meromyosin binding sites. The cross bridge sits poised and ready to bond to actin and create tension, but tropomyosin is in the way. When calcium arrives on the scene, it rapidly attaches to troponin; troponin then assumes a new shape and "levers" the tropomyosin out of the way, and cross bridge bonds are able to form. The ATP-carried energy becomes available, ATP becomes ADP, tension forms in the cross bridge, and the muscle contracts. The chain of events looks like this in schematic form:

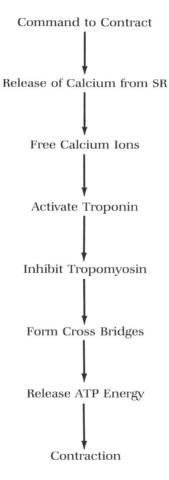

Command to Contract

Release of Calcium from SR

Free Calcium Ions

Activate Troponin

Inhibit Tropomyosin

Form Cross Bridges

Release ATP Energy

Contraction

Quite a sequence just to move a few proteins in a cell.

Let's now turn our attention to the top of this flow of events, to the source and events surrounding the command to contract.

THE COMMAND TO CONTRACT

So far we have worked our way backward from cross bridge tension to the command to contract at the cellular level. Rather than continuing to work backward from effect to cause, let's now reverse the flow of ideas and work from the ultimate control center, the brain, to the muscle.

The command to contract begins in the brain, and here the command can be a very complex set of biochemical and electrical events which are beyond the scope of this discussion. In the journey to the muscle, the command will change form several times, appearing alternately as an electrical or a chemical signal. To understand the connection between electrical and chemical signaling, it is best to examine each of these communication forms separately. The electrical signal is first.

In Chapter 1, we looked at cell membranes that were excitable. Nerves and muscles have excitable membranes, and these membranes have pumps that move specific ions in preferred directions; for example, they move sodium out of the cell and potassium into the cell. The membrane, however, is preferentially leaky, especially to potassium, and potassium tends to leak slowly out of the cell to the extracellular environment. At the same time, the membrane does not let any of the negative charges in the cell go with the leaking potassium ions. With positive charges leaving the cell and negative charges unable to follow, electrical neutrality inside the cell is lost. Electrical neutrality can be thought of as a condition in which the electrical charges, both positive and negative, inside a cell are equal in number. With potassium leaking out, the cell loses electrical neutrality, and the cell interior becomes charged to about 90 millivolts (90/1000ths of a volt) negative with respect to the cell exterior [17]. The curious situation set up by the combination of membrane-bound pumps and a preferentially leaking membrane potentiates one of the most important events in physiology—the action potential, which is the cell's electrical signal.

To understand more about the action potential, we first need to look at the response of a charged membrane when trying to remove its charge by applying small electrodes to a cell membrane with the electrodes connected to a battery. A battery voltage applied opposite to the membrane charge will reduce the effective voltage across the membrane. This is called *depolarization.*

Depolarizing a portion of an excitable membrane (the transmembrane potential moves toward zero), a set of electrical currents develops between the charged and uncharged parts of the membrane that causes a depolarization of adjacent membrane areas. If this depolarization proceeds past a certain threshold level, a complete irreversible

MUSCLE AND MOVEMENT

depolarization and repolarization cycle occurs. This sequence of depolarization and repolarization is the action potential. During the depolarization phase of the cycle, other adjacent areas of the membrane also depolarize to their threshold levels due to the currents flowing between the depolarized and still polarized portions of the membrane. With these currents depolarizing the membrane ahead of the membrane segment going through an action potential, the action potential is able to propagate along the membrane from the original site of depolarization.

The action potential is a communications datum within a single nerve or muscle cell. For most nerve systems, the cell transmits information within the frequency of the action potentials (the number of action potentials that occur per unit of time) when something stimulates the cell. The message travels to the next cell, however, by a chemical means—a neurotransmitter. The neuron makes and stores the neurotransmitter chemical, secreting it on command.

Generally, cells communicate with other cells using molecules secreted by one cell that migrate to another target cell to produce an effect. The use of molecules or transmitter chemicals has two cellular advantages over using the action potential alone for intercellular communications. First, the transmitter substance is secreted by the sending cell only and detected by the receiving cell only. Consequently, the information moves in only one direction, from transmitting cell to receiving cell. Second, limited chemical transmission has a high degree of communications security. The transmitter molecule can be quite specific and can fit only into unique receptors that are on the receiving cell membrane. Another molecule with the correct molecular shape of the transmitter molecule is the only other molecule that may be able to fit into the receptor. By placing the transmitting and receiving cells close together and using a chemical signal, neural communication flows in one direction only, with little cross talk to neighboring cells.

In contrast to a neurotransmitter, an action potential is not one-directional. With proper stimulation, an action potential can move in any direction on a nerve or muscle cell membrane. In addition, currents and potentials associated with action potentials can be picked up some distance from the original event. A good example is the heart; electrical events in and around the heart appear on arms, legs, and chest and are recorded in an electrocardiogram. In this sense, action potentials are less secure over relatively small distances than using chemical transmitters.

The relationships between neurotransmitters and action potentials may explain why they are alternated as the command travels from brain to muscle. The electrical events (action potential) are fast, providing a rapid propagation of information. A neurotransmitter, on the other hand, makes sure that the command stays in the proper channel, that the data flow is in the proper direction within the channel, and that all this introduces only a small time delay in the whole information transfer. With these facts in mind, let's return to the skeletal muscle system.

INTERPRETING THE COMMAND AT THE MUSCLE

For skeletal muscle, the electrical signal travels along the primary communicating cell, the motor neuron. The motor neuron secretes its neurotransmitter, *acetylcholine*, which initiates a muscle action potential when the acetylcholine attaches to the muscle membrane at a specialized portion of the muscle membrane called the *motor end plate.* Nerve and muscle are very close; so close, in fact, that only the electron microscope can show that they do not touch (Figure 2–8). This area of proximity and chemical transmission is called a *synapse.*

A muscle action potential, started by the attachment of acetylcholine to the membrane, makes its way into the cell interior through a fine network of tubules called *T-tubules.* These tubules are very small structures that are continuous with the exterior muscle membrane. In physical contact with the T-tubules *inside* the muscle cell are the special sacs *(terminal cisternae)* that store calcium (Figure 2–9). It is the action potential carried into the cell's interior by the T-tubules that causes the release of calcium from the SR. And free calcium ions initiated the sequence of events within the contractile proteins.

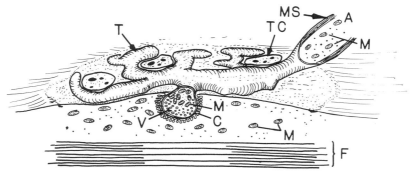

Figure 2.8. *Organization of the motor end plate. The motor end plate forms the connection between nerve tissue and the muscle cell. The nerve terminal (T) spreads over the muscle cell wall, interspersed with teloglial cells (TC). A mylelin sheath (MS) surrounds the nerve axon (A). Mitochondria (M) are located along the nerve terminal and axon, and are in the muscle cell. The muscle membrane has clefts (C) to receive the neurotransmitter, which is stored in the terminal vesicles (V), ready for transmission. F is the contractile proteins in the muscle cell.*

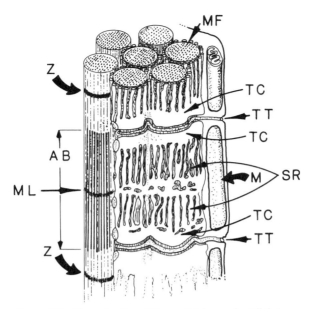

Figure 2.9. *Organization of the skeletal muscle cell for contraction. Calcium release and uptake is controlled by the sarcoplasmic reticuculum (SR) and the terminal cisternae (TC), which surround the individual myofibrils (MF). The action potential reaches the SR through the T-tubule system (TT) that connects to the outer cell membrane. Cell mitrochrondria (M) are located at the edge of contractile machinery. In mammalian muscle, the T-tubules are aligned with the edge of the A-band (AB) of the sarcomere. Z is the Z-line and ML is the Sarcomere M-line.*

We can now look at the entire sequence of alternating electrical and chemical events for a single muscle twitch. The command begins in the brain as a combination of electrochemical events. The signal comes out of the brain as an action potential on a neuron that transmits a chemical across a synapse to the motor neuron. Here a nerve action potential begins and propagates along the motor nerve to the muscle. At the nerve–muscle synapse, the nerve terminal releases acetylcholine, which travels to the muscle membrane and starts a muscle action potential. The muscle action potential propagates along the muscle membrane to the SR (via the T-tubules), which releases calcium ions (the last chemical transmitter) that bind to troponin. Troponin levers tropomyosin out of the way, cross bridges are formed, and tension develops. The muscle twitches, and this twitch is detected by intramuscular sensors that carry information about the twitch back to the brain. Considering all that is involved, it is astonishing that the system works at all, let alone that it works so well. With a muscle now contracting under control of the brain, let's take a look at some of the overall properties of contracting muscle.

PHYSICAL PROPERTIES
OF MUSCLE CONTRACTION

An idea introduced earlier in this discussion was that the actin-myosin array is very much like a crystal. And like a crystal, whose outer shape is a reflection of internal atomic associations, muscle function is a reflection of sarcomere function. A look at a few experiments on sarcomeres and muscles will illustrate this association.

Physiologists have successfully isolated the myofibril with its sarcomeres and measured the tension-building capabilities of these cell elements. Experiments disclose that a sarcomere's tension depends upon its length. The relationship between the length and tension appears like the curve shown in Figure 2–10 [18]. The tension under consideration is called *isometric tension,* that is, tension developed while the sarcomere is held at constant length (*iso* = constant, *metric* = length). If the sarcomere is free to contract when stimulated, the condition is called *isotonic* (*iso* = constant, *tonic* = tension) because

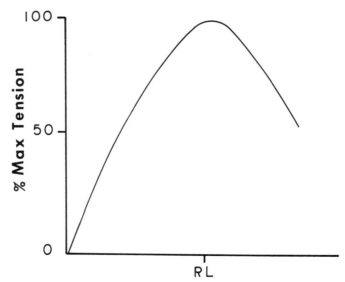

Figure 2.10. *Length-tension curve for a sarcomere. Because of the molecular arrangement within the sarcomere, it cannot develop unlimited tension. Tension is a function of sarcomere length. The maximum tension is at the resting length (RL) and decreases when the sarcomere is shortened (to the left of RL) and lengthened (to the right of RL).*

MUSCLE AND MOVEMENT

the developed tension is constant during the contraction [19]. One conclusion that can be drawn from the graph in Figure 2–9 is that the contractile machinery has a length at which it works best. If sarcomere length is either too long or too short, the tension it can develop decreases.

For the whole muscle, the length–tension curve looks like Figure 2–11. As the muscle becomes shorter than its resting length, the ability to develop tension decreases just like the sarcomere. This is true for a shortening muscle, but what if the length is greater than optimum? Experiments also show that stretching the muscle beyond its resting length creates first an overall decrease in muscle tension; then it suddenly begins to rise again [20]. Some mechanically sophisticated experiments provided an explanation for this increasing tension that follows the initial decrease.

Muscles are composed of cells held together with large amounts of connective tissue. Increasing muscle length beyond a certain point begins to stretch connective tissue elements in muscle. At the same time, the "active" muscle tension produced by all the sarcomeres follows the same curve produced by a single sarcomere. The sharp rise in passive tension seen in Figure 2–11 comes from the connective tissue component of the muscle. These passive tissue elements stretch as

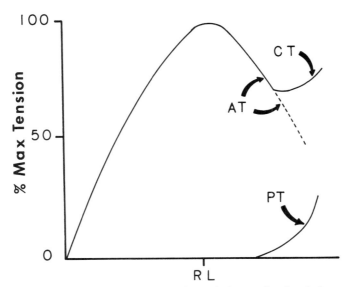

Figure 2.11. *Length-tension curve for a whole muscle. The whole muscle length-tension curve looks very much like the sarcomere. The greatest active tension (AT) appears at the whole muscle resting length. Beyond the resting length (RL) active tension (AT) decreases as the passive tension (PT) from connective tissue increases, and the combined tension (CT) thus increases.*

the muscle length increases, and because their response is nonlinear, the curve slopes upward very fast, rapidly contributing to the total tension measured in the muscle. The important element in the graph, however, is the active tension produced by the sarcomeres inside the muscle cell. The decreasing tension with increasing length for the sarcomere shown in Figure 2–10 appears also in the curve describing the whole muscle, and defines the tension-building capabilities of the whole muscle.

Considering the sequence of events we have built with protein filaments sliding past one another and cross bridges developing tension, how might this arrangement explain the length–tension curves? Let's consider first a sarcomere at its resting length as shown in Part A of Figure 2–12.

Lengthening the sarcomere beyond its resting length as shown in Part B of Figure 2–12 causes fewer opportunities for cross bridges to form because the actin and myosin overlap very little. The important difference is the amount of overlap between the actin and myosin.

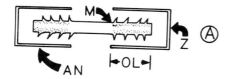

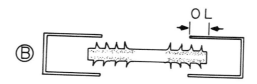

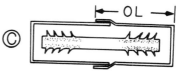

Figure 2.12. *Predicting the length-tension curve from the sliding filament model. A. At or near the resting length, the actin (AN) extends over the myosin (M) with the most effective overlap (OL). B. Extending the sarcomere length reduces the overlap (OL), and tension drops. C. Shortening the sarcomere too far causes too much overlap (OL), reducing tension again. Z is the sarcomere Z-line.*

Without cross bridges, a sarcomere cannot develop active tension for itself or for a muscle.

If the sarcomere is too short, as shown in Part C of Figure 2–12, the actin fibers begin to overlap. The overlapping filaments decrease the effective tension either by directly interfering with opposite cross bridge formation or by developing tension in the wrong direction at the overlapping segment. The exact mechanism is not clear, but the decrease in tension is quite real.

Muscle Properties in Exercise

On the practical side, the contracting capabilities of muscle shown in these curves flow from the basic protein organization. Exercise programs and systems like yoga work because of these subtle yet important molecular relationships. A closer look at yoga will show how this works.

A practitioner of yoga will develop muscle tone and strength through a series of what appear to be strange poses and muscle stretches. A closer look shows that many poses force muscle groups into extensions that allow only inefficient muscle contractions.

Yoga positions stretch muscles to lengths exceeding their resting length. Other poses and positions work on the other end of the muscle motion by making muscle groups too short to contract well. Forced to contract inefficiently, the moderate tensions typical of yoga easily stress the muscles. In turn, muscular activity stresses the muscle cell, causing it to increase in size and strength. But yoga is a somewhat static form of exercise. What about a more dynamic form of exercise?

Bicycling is a good example. Bicycle handbooks like to emphasize the need to raise the seat for the best possible cycling. Our newly learned muscle mechanics show why. Peddle-pushing comes from the extensor muscles of the legs, that is, those muscles that extend the leg against gravity. In order to extend the leg, the muscles must drape over the knee joint and connect to the lower leg. Because of this arrangement, small flexions of the knee quickly lengthen the extensor muscles beyond their resting length. When the leg is at or near full extension, an extensor muscle can develop its maximum tension. A seat too low will prevent full leg extension, producing an inefficient contraction of the extensor muscles. Thus, the easiest cycling comes from raising a bicycle seat and using the leg muscles efficiently.

Muscle Growth

Bicycling and exercise in general cause another process to occur: *muscular hypertrophy* or growth. Nearly all of us have experience with this growth as a consequence of exercise or just growing to an adult size. Clearly, the whole muscle increases in size, but the real question is: How does skeletal muscle handle this growth at a cellular level?

Understanding how muscle copes with growth requires first an examination of the controlling agent in the cell, the nucleus. When we first looked at a nucleus and the cell volume it controls (Chapter 1), it seemed that a relatively fixed ratio exists between the total cell volume and the amount of nucleic acids present in the cell. Looking at this ratio in growing muscle shows a nearly constant ratio, right up to rather large amounts of growth [21]. That could mean that either new cells were being made (called *hyperplasia*), or old cells were increasing in size without increasing in number (called *hypertrophy*) and acquiring nuclei from somewhere to keep the ratio constant.

Microscopic examinations clearly showed no increase in the number of muscle cells; instead muscle cells were increasing in size— hypertrophy. But at the same time, the number of nuclei seemed to be increasing during hypertrophy as well, keeping the cell-volume-to-nucleus ratio nearly constant. In some manner, the muscle cell is getting more nuclei—but from where? Nothing visually or chemically indicated that the cells were making new nuclei. If the cell does not make a new nucleus, it has no choice but to collect one from someplace else (cellularly speaking). Tracking events carefully showed that the muscle cell still retains its syncytium-making skills, incorporating a previously ignored cell called a *satellite cell* [22]. This cell hangs around just outside the muscle cell membrane staying close to things. With this sort of syncytial mechanism, the muscle cell takes little energy from the primary cellular activity of contracting to make more nuclei.

Do other uses exist for the satellite cell? Apparently not, for it hugs the skeletal muscle cell with an understudy's persistence, waiting for the time to be part of skeletal muscle growth. The arrangement seems to be an expression of: "Make them before you need them." What happens if no spare satellite cells remain? This appears in only rather extreme cases of muscular demand; when available satellite cells run out, new muscle cells start being made [23]. From all indications, this emergency hyperplasia is unusual and represents the heroic efforts of a system responding to a very severe demand. Thus, hypertrophy is the normal mode of muscle growth, presenting both an increase in cell size and an increase in intracellular protein.

Muscles grow in size when they are used. What then might be the effect of disuse? Perhaps the most familiar of these effects is the look of a limb after it is confined to a cast during the repair of a fracture. A comparison of a casted limb with its unconfined mate shows that disuse shrinks the muscles. In the casted limb, they are weak, small, and soft in comparison with the other, more active muscles. These effects of disuse are called *atrophy*. Examining the muscle cells microscopically at cast removal would show that they are small in diameter, with few mitochondria and only small amounts of contractile proteins languishing within the cell [24].

An example of disuse atrophy under more dramatic conditions is the muscle wasting that is associated with the prolonged weight-lessness of space. With no gravity to stress either the muscles or the bones, they both weaken. Special exercise programs are used by the astronauts to reverse this trend of atrophy. In addition, the astronauts perform regular experiments to determine the amount of atrophy in bone and muscle, and to determine if the exercise program is slowing or reversing the atrophy. The cardiovascular system also loses its conditioning to gravity, and severe low blood pressure is not uncommon when the astronaut returns to earth. The body quickly returns to the pre-flight condition in a few weeks, however. Fortunately, space-caused atrophy is reversible, which is a good example of whole-body adaptation to a changing environment.

For young and old alike, the use of muscle stimulates growth and strength. To support its functions, muscle in growth requires additional tissues—vessels to supply blood, nerves to control these new vascular beds, and connective tissue to collect and encapsulate the various cell aggregates [25].

On the coattails of hypertrophy comes the growth of many other tissue components. A prolonged muscle growth process seems also to leave a muscle "memory." This memory becomes evident when a muscle is allowed to atrophy after a prolonged stress and growth period. Later the muscle is again called on to hypertrophy. The muscle gains strength, tone, and size much faster the second time around than the first [21]. This more rapid second growth could be the result of supporting tissues (vessels, nerves, connective tissue) that were hanging around from the first growth and did not have to be remade. At the same time, we have to remember that the muscle cell is a living cell, with an ability to be "educated" no less than any other single cell organism. If a single cell entity like an amoeba can be taught rudimentary responses, then why not a living muscle cell? Whatever the reason, experience indicates that bodies made strong by exercise in childhood respond better to exercise in adulthood. Physical education in the young is building for the future. Without exercise, muscles weaken. The clear conclusion to this story is: Use it or lose it.

MUSCLE CRAMPS, TWITCHING, AND PARALYSIS

Up to now we have assumed that muscle movement was under control. Willing movement made it happen. But under certain conditions, this control is lost. Most of us have experienced muscle cramps or twitching, and many know the helplessness of paralysis. All these events (cramps, twitching, and paralysis) are examples of control-loss in muscle.

Understanding the elements of control-loss will require a review of the normal sequence of events in muscle contraction. The cascade of events for normal contraction is shown in Figure 2–13. The process can be aborted anywhere along the chain of events, but most interruptions occur at specific sites along the way. Let's look at some examples.

The first site open to manipulation is the relationship between the action potential and the membrane rest potential in muscle. This relationship is composed of a triad of interactions among the rest potential, the threshold to initiate an action potential, and free calcium *outside* the cell. Calcium, it turns out, has a function both inside and outside the muscle.

Placing an electrode inside a muscle cell and simply watching the rest potential over some length of time would soon show that the word "rest" is incorrect. A normal excitable membrane seems to experience all sorts of spontaneous depolarizations, but all too small to initiate an action potential [26]. These membrane potentials vary as

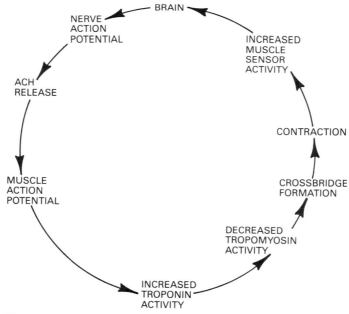

Figure 2.13. *The sequence of events and feedback during muscle contraction. The cycle of events starts and ends in the brain. The brain starts the command to contract and then must know what happened in response to the command. The information returns through the neural sensors within the muscle system.*

the condition of the cell membrane varies, affecting potassium and sodium permeability. The membrane seems to have a very dynamic sort of rest.

But if the threshold should move close to the rest potential, these spontaneous depolarizations could exceed the threshold and trigger a spontaneous twitch. Experiments show rather curiously that calcium outside the cell has an influence on the action potential threshold. A normal calcium level keeps the threshold electrically more positive, separated a healthy distance from the not-so-stable rest potential. When external calcium is lower than normal, the threshold moves to a more negative value, bringing the threshold and resting potential closer together [26]. Now the muscle experiences spontaneous action potentials that are not part of normal neural control. A whole series of rapid action potentials will produce a complete and massive contraction—a painful muscle cramp.

Changes in external calcium are only one way of altering these fine-tuned relationships. A change in temperature, metabolism, or fatigue can also produce cramps, a not-uncommon combination of conditions for the long-distance runner in the spring or fall. Because the events leading to twitches and cramps arise in the muscle, they cannot be "willed" away. Therapy requires restoring an equilibrium condition at the muscle to stop the spontaneous activities. But aside from these chemical changes, disease can also cause a loss of muscular control.

During the summers of the 1930's to the 1950's, a crippler of monumental proportions stalked the young in Western developed societies. It acquired several names, but the most familiar were *polio* or *infantile paralysis*. The cause was a virus, then unknown and unsuspected, that infected nerve cells, but not just any nerve cell. It preferred skeletal muscle motor nerves, the communicating link between brain and muscle [27]. When a motor nerve died from this viral infection, instructions to contract could no longer reach the muscle cells, leaving the muscles paralyzed. If the motor nerves to the diaphragm were destroyed, the victim would die without some mechanical assistance in breathing. Out of this grew the technology of the "iron lung."

Today, paralysis results less frequently from polio (thanks to immunization) and more often from trauma to the spinal cord. Whether by disease or trauma, the cause for paralysis is a lost neural link between brain and muscle.

Paralysis is neither always permanent nor always a disadvantage. Surgeons often use special pharmacalogical compounds to induce muscle relaxation or paralysis. These compounds usually work at the synapse between nerve and muscle where acetylcholine moves across the synapse. This site is also where the so-called nerve gases work.

In contrast to muscle relaxants that prevent acetylcholine from working at the muscle membrane by blocking acetylcholine attachment, nerve gases work by blocking the normal *inactivation* of acetylcholine by enzymes at the synapse. Inactivation occurs when an enzyme cuts the acetylcholine molecule in two. With no way of removing acetylcholine, the neurotransmitter accumulates in the synapse. At first it causes uncontrolled muscle contraction, then paralysis, finally death. Compounds like nerve gases are used as insecticides (organophosphates) and work because insects have a neuromuscular system chemically like ours.

Chemical poisons, disease, and trauma can all interrupt the command to contract on its way to the muscle. That we experience so few failures in the system speaks well for both its security and its reliability.

MUSCLE STRENGTH

With some idea of how muscles work, a good question would be: Just how well do muscles work?

Within the variability of expressing the human form is a range of muscle strengths that is just as large. In a way, physical strength is correlated with size; larger humans are generally stronger than smaller humans. Competitors in most strength-dependent sports are divided into specific body weight classes because of this fact. And even within these classes, muscles work in varying degrees.

On a weight-to-weight basis, other mammals show muscle strength seemingly greater than ours. Is this a fair comparison? Can we compare humans with animals or even insects? In making some comparisons, analysis requires considering that the muscle weight is part of what must be moved. Are insect muscles really stronger than a human's on a weight basis or is this just an illusion caused by the insect's smaller size and the special muscle arrangements available to an exoskeleton? Some cats, such as the leopard, are known to have unusual strength, and a 150-pound black bear is certainly stronger than a 150-pound man. Just why such differences occur is not entirely clear. The molecular events that occur in human muscle contraction occur also in insects and mammals in general. One important contributor may be the anatomical relationship between muscle and bone. A few examples of body levers in human anatomy may give some clues.

For most adults, it is not overly difficult to hold a ten-pound weight in one hand with the lower arm perpendicular to the upper arm. The arrangement looks like the diagram in Figure 2–14.

Redrawing this system into a set of levers as shown in Figure 2–15 brings out the essence of the problem.

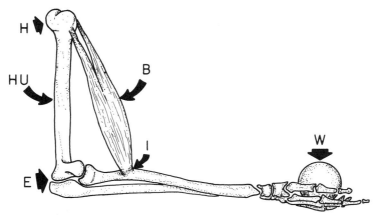

Figure 2.14. *The anatomical arrangement of bones and muscle in an arm holding a weight. The biceps (B) have two heads that connect to bone on the head (H) of the humerus (HU), and inserts on the radius (I). The fulcurm is at the elbow (E). W is the weight.*

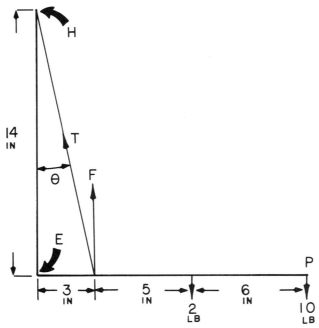

Figure 2.15. *The arm anatomy reduced to a lever system. H is the head of the humerus, E is the elbow, and P is the palm of the hand. Some typical values for an adult arm show that a tension (T) of 53 pounds is required to hold a ten pound weight at P. See the text for calculation details.*

The muscle handling the weight is the *biceps*, and the anatomy of this muscle's connection to the bone sets up the system. The bicep has two (*bi* = two) heads that insert on the humerus. The other end of the bicep connects to the lower arm about three inches from the elbow. E is the elbow point and also the fulcrum for the lower arm lever. The analysis is one of evaluating an equilibrium condition in which the sum of torques tending to move the weight upward is equal to the sum of torques tending to move the weight downward. The calculations are not too complicated, provided we work through the equation with one eye on the solution.

The forces trying to move the arm downward are: (1) the weight of the arm (two pounds applied at the center of gravity), and (2) the ten-pound weight in the palm, *P*. The upward force comes from the muscle labeled force, *F*. Torque is equal to force times the length of the perpendicular lever arm. The sum of torques looks like:

$$T = (F \times 3) - (2 \times 8) - (10 \times 14)$$

The total torque, *T*, is zero with equilibrium, and nothing moving.

$$0 = 3F - 16 - 140$$
$$0 = 3F - 156$$
$$F = 156/3 = 52 \text{ pounds}$$

Thus, the biceps produce an upward force of 52 pounds to hold that ten-pound weight. The muscle, however, is not acting vertically. It has to develop even more force within itself because of the muscle-bone geometry. The tension, *F*, developed in the muscle is:

$$S = F/\text{Cos } a, a = 12 \text{ degrees}$$
$$S = F/\text{Cos } 12 \text{ degrees}$$
$$S = 52/0.978 = 53.2 \text{ pounds}$$

Thus, this muscle must develop 53.2 pounds to hold that ten-pound weight. In a practical sense, a weight-lifter holding 100 pounds in that position would require the biceps to generate 532 pounds of force— over a quarter of a ton. It is time to appreciate those microscopic sarcomeres. Nevertheless, the question remains: Why should muscles inefficiently use 53 pounds to lift ten pounds?

In a technological society such as ours, efficiency is important from both academic and practical points of view. The technical answer to our question is easy to find. Muscles drape over an inefficient lever system because of the position of the fulcrum at the elbow joint and the forces around that joint. All the forces are on the same side of the fulcrum with muscle tension acting through a shorter lever-arm than the weight.

The arm is only one example. Another resides in the jaw with the muscle *(masseter)* on the same side of the joint *(fulcrum)* as the teeth. Still another example is the lever in the feet and ankles. The calf muscle, for instance, must develop 600 pounds of tension to hold a 150-pound man on his toes. Obviously the problem is not the muscle strength itself—it is the form of the anatomical levers.

This is part of the muscle inefficiency question, certainly the physical aspect of it. Still, it is tempting to ask again, why—what is the gain with this sort of inefficiency? Earlier we noted the fine movement of the body, a hallmark of human function. The levers that make up the human body are capable of very fine movement. That short lever arm combines with the muscle to smooth out the jumps that normally occur in muscle contraction. Small changes in tension that are characteristic of muscles in isometric tension are often too small to be seen outright. It appears that humans traded strength for fine movement—perhaps not a bad trade after all. Mankind with stronger levers might have changed the course of human development. The fine movements required in the practice of art, music, technology, and medicine might not have been possible. Human cultures and ways of life are in large part a product of human anatomy and physiology.

CONCLUSIONS

For most of us, movement is simple. Will a movement and it happens. Underlying this movement, however, is some sophisticated biochemistry, utilizing both mechanical orientations of proteins and an elaborate set of controls to regulate events.

Muscle, as a living tissue, responds to stress with growth to improve both muscular strength and efficiency, and seems to learn from past experiences in stress and growth.

The sliding filament model for muscle contraction is a geometry carried over to contractile proteins found in other living things. Contractile machinery resides in single-cell animals like the amoeba and appears in highly developed forms in more complex sea animals and insects. Sliding filaments even seem to play a role in the division of DNA into daughter cells during cell division. The sliding filament model is both an elegant explanation of our many observations about muscle contraction and a successful means of predicting future experimental results.

3

THE LIVING PUMP
AND ITS FLEXIBLE
PIPES

Unlike the liver or the spleen, the heart does its work with a perceivable rhythm. It is a dynamic organ vibrating the chest wall with each beat. It is alive, composed of living cells that contract with each beat. When we run or exercise, our heart rate increases. How does the heart know to do this? Even after we eat a heavy meal, the heart seems to beat more strongly. Why? How does the heart know that it is functioning properly for body needs? And what are these needs? These and many other questions will be answered as we tour the cardiovascular system.

THE CARDIOVASCULAR PATTERN

To move oxygen, nutrients, and waste products throughout the body, we have a circulatory pattern that forms a closed volume. The energy to propel blood through our circulation comes from the heart. The heart is central to the circulatory pattern, and normally everything that flows through the blood vessels must also flow through the heart. The physical contractions of the heart supply energy to push blood through the great array of flexible arteries and veins. Despite its central position in things, the heart is not the only pump in the body. A second pump, called the skeletal muscle pump, also actively moves blood through the circulation. The role of this second pump is less obvious, but under conditions that force it to fail, its importance quickly emerges. This pump will be described in more detail later.

Fanning outward from the heart are the pipes that channel blood moving to and from the heart. The vessels are named according to the direction of blood flow relative to the heart. *Veins* carry blood toward the heart; *arteries* carry blood away from the heart. In the body, occupying the separation between arteries and veins are the *capillaries*, a network of extremely small vessels, which are the sites of gas exchange between the cells in the blood and the interior of the body. Gas exchange, however, is only one of the many events that occur in capillaries.

The heart functionally and anatomically divides into two parts, called the right and left heart, as shown in Figure 3–1. This is really an overly simple division, for the "right heart" is not limited to the right side of the body in either the vascular beds it serves or its physical location. Each side is really a separate pump that shares common walls in both the ventricles and the atria. The two pumps, however, have quite different pressure and work requirements. The right ventricle pumps at a much lower pressure than the left ventricle and does not work as hard. The right ventricle pumps blood into the pulmonary circulation (lungs), which is a low-pressure, low-resistance system. Within the lungs, the capillaries are very small and delicate with little

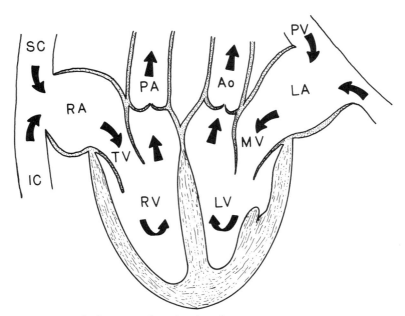

Figure 3.1. *The heart as a four-chambered structure. SC, superior vena cava; IC, inferior vena cava; RA, right atrium; TV, tricuspid valve; RV, right ventricle; PA, pulmonary artery; PV, pulmonary veins; LA, left atrium; MV, mitral valve; LV, left ventricle; Ao, aorta. Although the anatomy of this drawing is not entirely correct, the flow sequence is, which shows the distinct four chambers of the heart. Each atrium and ventricle makes a two-stage pump with separate input and output channels.*

supporting tissue. Along with the capillaries is a well-developed network of lymphatic channels.

The lung capillaries are designed to exchange gasses through a membrane that measures less than 1/1000th of a millimeter thick and separates the blood from the air in the lung. During the gas exchange, oxygen moves toward the blood, while carbon dioxide moves from the blood into the lung air.

In contrast to the right ventricle, the left ventricle pumps into the whole-body circulation, which is a high-pressure, high-resistance system. When the blood finally reaches capillary beds in the body, another gas exchange occurs. Oxygen diffuses out of the blood into the tissues and carbon dioxide diffuses out of the tissues back into the blood. Each side of the heart, then, pumps into a system of pipes and channels that determine the work level for each side of the heart.

Let's pull it all together now and look at the total flow pattern. The circulatory pattern is: venous blood, low in oxygen and high in carbon dioxide, flows into the *right atrium*, then into the *right ventricle* through the *tricuspid valve;* the right ventricle pumps the blood through the *pulmonary artery* (it is an artery because flow is away from the heart); in the lungs, oxygen moves in, carbon dioxide moves out, and the blood flows to the *left atrium* via the *pulmonary vein* (it is a vein because flow is toward the heart); from the left atrium to the *left ventricle,* the blood flows through the *mitral* or *bicuspid* valve; the left ventricle pumps the blood into the *aorta* through the *aortic valve;* the blood flows to the body through the arteries and finally to the capillaries; in the capillaries the blood gives up oxygen to the tissues, takes up carbon dioxide from the tissues and becomes venous blood to complete the circuit.

With the flow sequence laid out, this is a good time to consider the character of the transported material.

A CAPSULE VIEW OF BLOOD

Moved about within the circulation is a rather complex mix of materials called blood. Because blood tends to flow freely from a wound, it is hard to think of it as a tissue. Except for its fluid qualities, however, blood satisfies all the requirements for being one. Like any tissue that has a supporting material to hold cells together, blood has a suspending medium called *plasma.* In whole blood are: a large number of proteins that participate in the immune response, that function as hormone carriers, and that are sometimes sources of hormones; *white blood cells,* called *leukocytes;* small, colorless corpuscles called *platelets;* and an astronomical number of small discoid cells without a nucleus, called *red blood cells.* This list is very general and incomplete, but adequate to appreciate the complexity of the blood flowing within the vascular compartments.

Blood in the animal world is not always red. Blood can also be green, purple, or blue, depending upon the pigment used to carry oxygen [1]. Ours is red. The color of human blood comes from the red blood cell, which contains an oxygen-carrying protein called *hemoglobin*. The oxygen binds to an iron molecule suspended in a molecular ring inside hemoglobin. Through some very subtle molecular manipulating by the molecular rings, the oxygen is bound to the hemoglobin, allowing the red cell to transport oxygen to the tissues [2].

Unlike oxygen, carbon dioxide is transported to the lungs outside the red cell. It arrives at the lungs in the plasma as a bicarbonate ion (the chemical formula is $-HCO_3$). Nevertheless, the red cell does carry the enzyme that speeds the chemical reaction of carbon dioxide and water to form carbonic acid, which separates into a bicarbonate ion and a hydrogen ion in the plasma. Any symmetry we might have expected in the transport of oxygen and carbon dioxide does not exist.

Now that we've looked at the overall circulatory pattern and learned a little about the circulating blood, let's focus in on a few individual circulatory components. The starting point is the site of some dynamic events, the vascular bed.

THE VASCULAR BED

Once blood is set into motion by the heart, there remains the very practical business of getting it close to the tissues and cells needing the materials carried by blood. It is the vascular bed that carries out this final distribution. Figure 3–2 shows the circulatory pattern in a form we can use to better understand the organization of the vascular bed.

The first relationship to draw from Figure 3–2 is that vascular beds are sometimes in parallel and sometimes in series. The series beds, however, are at the same time in parallel with the remaining circulation. This parallel configuration for the whole system offers some unusual properties. For example, this arrangement lets each vascular bed regulate its own flow with relative independence of the others. Flow through the various beds begins to diminish only when the heart's ability to maintain an adequate blood pressure begins to decrease, or when the total flow required for all the vascular beds exceeds the heart's pumping capacity. Diseases that damage the heart directly can greatly reduce its abilty to keep an adequate flow. Independent control of flow at each vascular bed lets the circulation meet the special and changing needs of each tissue.

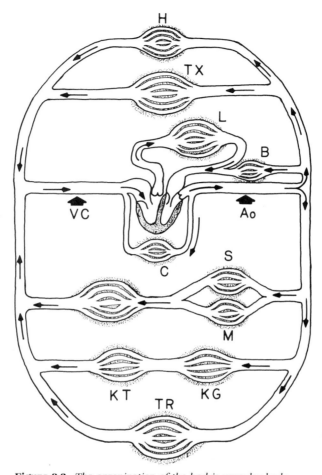

Figure 3.2. *The organization of the body's vascular beds. Blood leaves the heart through the aorta (Ao), and returns through the vena cava (VC). The vascular beds are: H, head; TX, thoracic; L, pulmonary; B, bronchial; C, coronary; S, splenic; M, messenteric, which collect together and pass through the portal; KG, renal glomerular; KT, renal tubular; TR, trunk and legs. All the vascular beds are either in parallel or series-parallel to permit each vascular bed to handle its own flow independently of the others.*

The complexity of a vascular system is hard to grasp from a diagram. In turn, if the diagram were completely accurate, the complexity of the drawing would obscure any appreciation we might have for arterial branching, divergence to capillaries, and convergence back to a single vein on the other side. One way to gain this appreciation is by looking at the division and the number of vessel types found in a well-studied system, in this case, the dog's mesenteric vascular bed [3].

The *mesenteric vascular bed* is used to collect materials absorbed through the gastrointestinal tract and transport these materials to the rest of the circulation. The divergence and convergence process starts with the aorta and ends with the *vena cava*. Table 1 shows each vessel division and the approximate number of vessels of each kind.

We can draw several observations from Table 1. One of the most obvious is the symmetry between corresponding segments on the arterial and venous sides (large arteries versus large veins, main artery branches versus main venous branches, etc.). Another facet is the rapid branching from the terminal branches to arterioles. The corresponding convergence from venules to terminal veins repeats this geometry, but in reverse. Additionally, at each branching level, the vessels are all in parallel with one another. Thus, those 1,200,000,000 capillaries are all in parallel.

Molecular transfer occurs at the capillary level for both gasses and nutritional molecules, and to support this transfer, the capillaries in this particular system provide an effective transfer area of about 600 square centimeters (about 93 square inches) [4]. This area is quite small when compared to the transfer area found in the lung (70 square meters) or gastrointestinal tract (300 square meters). We are, however, looking at only one small portion of the total circulatory system. If we include all the vascular beds in the body, the transfer area increases to a much larger value.

Table 1 divides the various vessels into 11 kinds. The division is based not so much on branching processes as such, but more on the microscopic qualities of the vessel walls. The larger arteries and arterioles (the last smooth-muscled vessels that enter a vascular bed) have relatively thick vascular walls that contain large amounts of smooth muscle. These vessels are not involved in the process of molecular

TABLE 1

KIND OF VESSEL	NUMBER OF VESSELS
Aorta	1
Large arteries	40
Main arterial branches	600
Terminal branches	1800
Arterioles	40,000,000
Capillaries	1,200,000,000
Venules	80,000,000
Terminal veins	1800
Main venous branches	600
Large veins	40
Vena cava	1

transfer to the tissues; their walls are too thick. Instead they serve to move and regulate blood flow. The smooth muscle in a vessel wall lets the vessel change its size, and with that size change, affects blood pressure, blood flow, and blood volume both within the vessel and in the capillary bed.

The capillary bed is filled with very fine vessels, often no more than eight microns (8×10^{-6} meters or about 3/10,000ths of an inch) in diameter. The red cell is about the same size. Thus, as the red blood cells enter the capillaries, they pass close to the cells they ultimately support, often only a single-cell thickness away.

Venules, like arterioles, are small-diameter vessels, but do not have enough smooth muscle to change their diameter significantly. The veins collect the blood that passes through the capillaries and compose a low-pressure, low-resistance vascular system. Some of the larger veins have some musculature, but only in limited amounts [5]. Through this framework, then, blood moves, giving life to cells located quite far from the surface of the body.

Along with oxygen, other materials that are essential for life also exchange at the capillary. Because most of the transfers at a capillary depend upon a diffusion process, the capillaries improve the limited cellular diffusion area by forcing the red cells to move in single file. Micrographs show that in many capillaries, the red cell is squeezed and deformed as the capillary diameter becomes slightly less than the diameter of the red cell. In these thin-walled vessels, the red cell's gas-carrying machinery gives up oxygen and supplies the enzymatic system to form the carbonic acid for carbon dioxide transport. Oxygen is not "squeezed" out of the red cell; rather, the intimate contact between cell and capillary wall ensures very short distances for diffusing oxygen and carbon dioxide down concentration gradients.

Both in the body and in the lung, capillaries work in very much the same way. The driving force for the gas exchange in the tissues and in the lungs is the difference in molecular concentration between the environment outside and inside the vascular compartment. The movement is from a site of high concentration to a site of lower concentration, so, if the oxygen concentration is higher in the vessel, the oxygen movement will be down its concentration gradient to the outside.

Experimental evidence indicates that nearly all the important molecules move into and out of the capillary by diffusion [6]. Electronmicrographs, however, suggest that other mechanisms may play a role in getting things into and out of the capillary. For example, molecular pumps within the cell membrane may move molecules through the capillary wall. Another process called *pinocytosis* is used by the cell to ingest rather large-sized objects. The cell does this by surrounding the object with some of its exterior, pinching off the resulting invagination, and moving it to the other side of the cell to once more combine with the membrane for release [7]. But some capillaries exercise a great deal of control over the free movement of materials that pass through their walls. The brain, for instance, puts up a well-defined barrier to certain molecules designated for exclusion from the brain. These capillaries form the so called blood–brain barrier.

Because a large body locates the cells far from the outside world and oxygen, the body uses an elaborate network of small vessels to get things to and from the individual cells. A failure in this supporting system results in cell death. Thus, if disease, injury, or blood clots reduce flow into a vascular bed, the tissues supported by those vessels suffer. If the flow stops completely, this produces a region of cell death called an *infarction*. If the infarction happens to be in the brain or the heart, the victim can quickly die or be left crippled.

Blood flow into capillaries depends on the flow through the arteries, which turn out to be very dynamic channels. They deserve a closer look.

THE ARTERIES AND THEIR CONTROL

The arteries regulate flow into the various vascular beds. These vessels range in size from the large vessels such as the aorta down to the small *arterioles* that have the last remnants of any smooth muscle. This smooth muscle is a physical means of providing control of blood flow through the artery.

Two other structures also influence the flow of blood into vascular beds. The first structure appears in some capillary beds as a heavily muscled vessel that directly connects between the artery and vein, providing a channel that can bypass the whole vascular bed. This structure is called an *arteriovenous* (A-V) *shunt* because it shunts blood flow past the capillaries. When the shunt is open, a portion of the blood that would normally reach the capillaries is, instead, routed past the vascular bed, never reaching the capillary network.

The second structure that changes flow into the capillary bed is a small circular band of smooth muscle in the arteriole called a *sphincter*. A muscular sphincter opens and closes on local command, setting up very tightly defined regional flow patterns. With these two controls, the arteriovenous shunt and sphincter, the flow pattern through a vascular bed and the commands that influence each controller become rather complicated. A schematic diagram of a vascular bed is shown in Figure 3–3.

In front of the shunts and sphincters are the arteries, however. And because of the mechanics of fluids, the arterial lumen size is a very effective control of both the amount of blood that reaches the

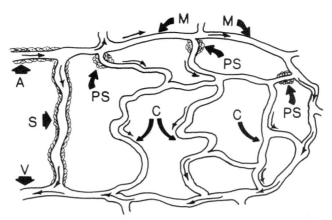

Figure 3.3. *Organization of a vascular bed. Flow through a vascular bed is controlled by several, often independent, factors. Flow can bypass the bed through an A-V shunt (S) and arteriole (A) drains directly into the venule (V). The shunt is under neural control. Blood is fed to the capillaries (C) through the metarteriole (M). Volume and direction of flow in the capillaries are controlled by the precapillary sphincters (PS). The result is a very sophisticated control of blood flow through the capillary bed.*

capillaries and the blood pressure exerted on the capillaries. (The lumen of a vessel is its functional internal opening.) Capillaries are fragile structures and will passively respond to their internal pressure. Controlling blood pressure at the capillaries keeps them safe from pressure-caused damage. Capillary destruction, for example is not an uncommon consequence of uncontrolled hypertension (high blood pressure).

To gain a handle on the principles involved, let's take some instruction from common experiences of trying to get fluids through various pipes like a garden hose or a soda straw.

Long, thin garden hoses, for example, do not deliver large volumes of water, and moving a milkshake through a straw can be very demanding at times. These experiences offer some good ideas about what controls fluids traveling through pipes. Increasing the thickness (viscosity) of a fluid will slow its flow through a pipe, and the job gets tougher with smaller pipes (drinking a thick milkshake through a straw can be tough work). In general, a small diameter in the pipe will reduce fluid flow. And making a pipe longer will also slow the flow, because a longer pipe reduces the pressure near the output end of the pipe.

Fluids move down the pipe's energy gradient, that is, from higher energy (pressure) to lower energy. Thus, if the difference in pressure along a pipe drops too quickly, flow will be reduced. All these factors fit together into a single equation that describes the elements that control flow, which is:

$$F = \frac{(P_1 - P_0)r^4}{8nL}$$

Let's see if everything is present in this equation. An increased pressure difference between the input pressure (P_1) and the output pressure (P_0) will increase flow; a decrease will reduce flow. Increasing the fluid viscosity (n) will decrease flow. And if the length of pipe (L) should increase, the flow (F) will decrease. Changing the lumen size has a very dramatic effect, with flow changing as the fourth power of the lumen radius (r). And because of this power factor of four, changing a lumen size is a very sensitive way of controlling flow. Now all the smooth muscle in arterial walls makes sense.

We now know that arteries have a great deal of smooth muscle, and this muscle is used to control the size of the artery. But how does the artery know when to change its size? A short answer to that question is: the brain tells it. But that answer is too short. The brain is a long way from the arteries located in the body, so control must be handled through some sort of communications link, which will be nerves—at least to begin with.

Smooth muscle, whether in vessels or other organs, is generally controlled by nerves coming from a segment of the nervous system called the *autonomic nervous system*. This system is further divided into the subgroups of the *sympathetic* and *parasympathetic* systems. The part of primary interest is the sympathetic system because it is the major controller of arterial size. The parasympathetic portion also has cardiovascular influences, and they will be brought into the discussion a little later. For now, one primary target is the sympathetic system and the neurotransmitter these nerves use to communicate with the smooth muscle cells.

The sympathetic nervous system uses a neurotransmitter that has two names, *norepinephrine* or *noradrenaline*. The compound has a shape quite like adrenaline secreted by the adrenal gland. This similarity in shape will crop up as a key feature later in the discussion.

The sympathetic nerves signal the smooth muscle in the vessel walls to contract by secreting norepinephrine, which binds to the smooth muscle membrane and stimulates contraction. As the level of nervous activity increases, the amount of released norepinephrine increases, recruiting even more smooth muscle cells and increasing their contraction, closing the vessel lumen further. This seems to be a one-way control of arterial narrowing, following the activity of the sympathetic nervous system.

Even so, to control the arteries properly, the autonomic nervous system must "know" what is going on in the arteries. To acquire this information, sensors throughout the cardiovascular system "feel" what is happening inside the vessels. Although information comes to the brain from pressure sensors throughout the vascular system, two major sites provide most of the details about pressure events in the main arteries.

The first site for pressure sensing is the *baroreceptors* located at the bifurcation of the carotid artery, where the artery divides into the internal and external carotid arteries. The sensors are made of a lacework of fine nerve terminals that penetrate deep into the arterial wall [8]. As blood pressure stretches the walls of the artery, the nerve terminals also stretch, and nerve activity increases with increasing pressure. In this way, the brain knows how much pressure is being applied to the carotid artery through the frequency of action potentials sent from the baroreceptors.

The second major baroreceptor location is in the walls of the *transverse aorta*, where this vessel changes direction and begins to head for the lower body. The locations of these two baroreceptors are shown in Figure 3–4.

The nerve terminals that make up the baroreceptors work like strain gauges, responding to the physical stretch of the arterial wall with changing internal blood pressure. Because the sensors are physical strain gauges, they can respond just as well to physical manipulation from the outside. For example, the normal arrival of blood pressure to one of the carotid sensors will stop when the common carotid artery between the heart and the baroreceptor is clamped off. When this happens, the arterial walls at the baroreceptor collapse, and nerve terminals at this receptor location tell the brain that the blood pressure has dramatically decreased. In response, the sympathetic nervous system increases activity, major arterial lumens actively narrow and the heart rate increases, all to restore the blood pressure. And because this stimulation of the nervous system is total, the combination of narrow vessels and an increased heart rate causes a much-higher-than-normal blood pressure in the arteries and heart chambers. Mechanically reducing pressure to one baroreceptor fools the vascular control system into believing that the pressure is too low, when in fact it is not. This experiment is like putting ice on a thermostat that controls the heater to convince the system that the room temperature is too low.

The system can respond in the other direction as well when pressure is applied to the baroreceptors from outside the body. Increasing physical pressure at a baroreceptor makes the vascular system respond with a rapid decrease in heart rate and arterial dilation, rapidly reducing the systemic blood pressure. If the blood pressure becomes too low, blood flow to the brain may not be enough to keep the brain conscious, and the victim faints. Some so-called "choke holds" that are part of the martial arts apply external pressure directly to the carotid artery baroreceptors, inducing a rapid cardiovascular

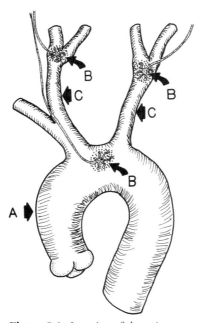

Figure 3.4. *Location of the primary baroreceptors in the central circulation. Baroreceptors (B) are sensors (nerve endings) arranged to detect changes in vascular diameter resulting from changes in systemic blood pressure. Other sensors are distributed throughout key vascular beds, but only these sensors have a major influence on systemic blood pressure. A is the aorta; C are the common carotid arteries.*

response and fainting. In general, then, the baroreceptors supply sensory information about the systemic blood pressure by tracking stress to the arterial walls from physical manipulation, from either outside the body or inside the vessels.

The brain's response to blood pressure is carried to the vessels and the heart by two routes, nerves and hormones. And it turns out that the substances used to convey the messages by each route are quite similar.

An autonomic nevous system activated by rage, fear, or excitement causes the *adrenal medulla*, within the adrenal gland, to secrete large amounts of *epinephrine* or *adrenaline*. Although epinephrine and *norepinephrine* are nearly alike in molecular shape, they have slightly different systemic effects as the body sorts out the response according to the shape of the molecule. For example, norepinephrine causes the arteries to narrow and increases the heart rate at the same time. On the other hand, epinephrine also increases the heart rate, but at the same time, it opens large vascular beds that supply blood to the skeletal muscles. These activities of epinephrine and norepinephrine prepare the body to either fight or take flight, and the whole process has been coined the "fight-or-flight response." Whether on a moment-to-moment basis for control of the blood pressure or to prepare the body for fight-or-flight, the baroreceptors and the sympathetic nervous system provide the adjustments that match the cardiovascular system to the body's needs.

Autonomic blood pressure adjustments can go astray at times, however. When this occurs a person can experience either high blood pressure *(hypertension)* or low blood pressure *(hypotension)*. We will later look at these abnormal pressures and some of the medical therapy used to restore normal blood pressure.

With some idea of events on the arterial side of the vascular bed, let's take a look at the venous side.

THE VEINS

Compared to the arteries, veins certainly look undramatic. At the outset, they have little or no musculature and generally have larger diameters than the corresponding arteries. The veins seem to receive only small amounts of nerve tissue from the autonomic nervous system. On the whole, the venous system is a low-pressure, low-resistance system. Still, veins play a surprisingly dynamic role in adjusting the whole circulation. Veins are more than just a passive collection depot for oxygen-poor blood.

First, veins may have a low blood pressure, but the volume of blood they transport over time must be equal to the arterial blood flow. Although this may not hold true for short time periods, it must in the long term. Conservation of mass sets the conditions for flow through the whole system. Without a flow balance, blood would pool or be depleted in various segments of the circulation. During normal periods of vascular transition from one state to another, pooling into special segments of the circulation does occur. If the pooling is uncontrolled or not eventually corrected, the results can range from hardly noticeable to the very obvious.

The first clue to the role of veins in the circulation comes from an uneven distribution of blood between arteries and veins. Depending upon the vascular bed under consideration, 55 to 80 percent of the blood is pooled on the venous side of the circulation [9]. Moreover, that distribution can change when blood pressure, arterial dimensions, or venous dimensions change. The small amount of smooth musculature found in some large veins may have a specific role here. Small changes in the diameter of a larger vein will change the volume of blood it can hold.

If the venous capacity (the vascular volume available for storing blood) in a vascular bed were to rapidly increase, blood returning to the heart from that bed would decrease to fill the expanding venous volume. Eventually the flow would again balance because the venous capacity cannot increase forever. The amount of blood returning to the heart can go almost to zero if a venous change in volume is both rapid and large. In turn, reducing blood flow to the heart will reduce flow out of the heart to the brain, leading to dizziness or fainting. Clearly, a circulation using capillaries that can engorge, arteries that can change diameter, strategic arteriovenous shunts, and venous capacitance vessels is a circulation able to distribute, redistribute, and fine-trim the blood volumes found in arteries, capillaries, and veins, all according to the body's needs.

Because veins make up a low-pressure system, often with a low driving energy inside thin, distensible veins, blood returning against the force of gravity from the lower limbs can run into difficulties. To help this return of blood to the heart, many of the veins larger than the venule have valves distributed along their lengths. The valves are made of thin cusps that open when the blood attempts to move away from the heart. The cusps themselves are made of connective tissue, and like other parts of the circulation, are covered with endothelial cells that do not stimulate blood clotting. These venous valves will have a central role in the second, muscular pump that is part of our circulation.

Both veins and arteries have autonomic innervation, but below the arteriole and the venule level, the vascular bed appears to have no neural input. Control at this level is contained within the vascular bed. These controls display some curious properties that deserve additional attention.

LOCAL CONTROL OF CIRCULATION

Although a large portion of the arterial system is innervated by the sympathetic nervous system, motor nerves do not appear to innervate the vessels below the arterial level, but do innervate the arteriovenous shunt [10]. This leaves a large part of the local circulatory pattern under a different sort of control. Just how this regulation takes place in nerve-free vessels is not entirely clear, but lacking nerves, an obvious choice is a hormone or several hormones. In some manner, the local tissue needs influence the vascular bed providing instructions on the amount of flow the bed must receive.

Because the vascular bed is under local control, some unusual situations can arise that would otherwise have little chance of happening. For example, with all efforts to conserve fuel, many homes and offices operate at temperatures of 68 degrees (F) or less. At higher room temperatures, the body is not called upon to exercise all its options to keep the internal body temperature normal. With lower environmental temperatures, capillaries close to the skin surface have a conflict to resolve: they must receive enough blood to keep the dependent cells alive, but also receive as little blood as possible to conserve body heat. A means of handling this is to have the blood flow oscillate into a vascular bed. Blood flow to outer portions of the hands appears in a cyclic fashion that is quite independent of any neural influences, swinging from limited flow to heavy flow [11]. If the cycle times for the two hands are different (and the cycles can be different because each control is independent), we can experience times when one hand is cold to the touch while the other is warm. On a more organized basis, temperature sensors placed on various parts of the body show a cyclic process of flow control as each independent vascular bed responds to the local flow needs [12].

Smooth muscle sphincters help set up this local control, and flow patterns are sensitive not only to the presence of several different ions, but also to the local temperature [12]. Some circulation beds like those found in the brain respond to oxygen and carbon dioxide levels in the blood, independent of attempts to interfere using drugs or manipulations of the autonomic nervous system. This sort of independence suggests the body's use of hormones for coordinating events on a very local level.

A hormone could be described as a substance released by tissue A that travels to tissue B, acts on the cells in tissue B, and then disappears with local metabolism. Biochemists have found two general molecular species that could provide such local hormonal activities. One hormone is part of a group of substances called *polyamines* [13]; the other is part of another group called *prostaglandins* [14].

One polyamine many people are familiar with is *histamine*, which can quickly affect a very large number of vascular beds, especially in the nose, producing the stuffy nose of an allergic response or viral infection. Histamine appears when cells respond to an allergen or to cell damage caused by an infecting virus.

Prostaglandins, on the other hand, are recent discoveries and seem to be used nearly everywhere in the body. They work on a vessel's smooth muscle directly, and also seem to modulate responses of the sympathetic nervous system innervating the arterial portion of the circulation [14]. Prostaglandins are very transient entities, flickering into existence for only a few seconds' time, doing their job, and just as quickly vanishing. Such hormones are hard to find, hard to concentrate for analysis, and hard to understand as controllers.

Confounding the fact that activities inside a vascular bed are regulated to a major degree, hormonal regulation often looks like a retrograde control, that is, with conditions *downstream* determining what is happening *upstream*. Just how this sort of retrograde control happens is yet unclear. The difficulty becomes most evident if we just consider how events downstream of a sphincter could get information back upstream to tell a sphincter when to close. Events here are still poorly understood.

So far, we have looked at each of the circulation elements that form the circulation pattern, except for the source of energy to move the blood. The heart is next.

THE HEART

Central to the circulation pattern, providing the kinetic energy to propel blood through the body's miles of vessels and generating sounds that radiate over the surface of the chest is the heart. It is a living pump made of cells specialized for contracting, conveying electrical signals, and making a connective tissue skeleton or framework to hang muscle, vessels, and valves upon. Dominating the cell population is muscle, either as a major source of contraction or as the major means of conducting electrical events to coordinate contraction. The heart's pumping action comes from a combination of specialized anatomy, electrical events, and mechanical shortening.

The heart's anatomy has been known for over two thousand years, but this knowledge of itself provided no information on how the circulation worked. Without the concept of a closed circulation, the heart as a pump made no sense. So, it became the "seat of the soul." Valves were thought either to control the flow of air or to prevent air from entering the heart's chambers [15]. In fact, it was not entirely clear why the heart needed chambers at all. Until the heart was examined using the technologies of this century, its workings were not understood.

Today we know the heart as a four-chambered pump with valves to keep blood moving through the heart in one direction. Earlier, Figure 3–1 showed the heart's chambers laid out in a very non-anatomical way. We used the figure only to show the four-chambered configuration. Figure 3–5, on the other hand, gives quite another view of the heart.

In outline, the heart takes on the general form of a smooth inverted pyramid with its base toward the head. Connected at the heart's base are the great vessels. The apex of the heart points downward and to

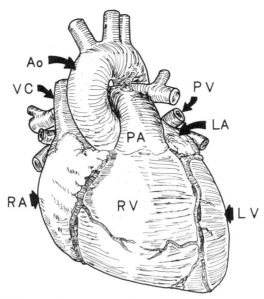

Figure 3.5. *An anatomical profile of the heart. This diagram provides a truer anatomical picture of the heart than Figure 3.1. The heart is like an inverted pyramid with its apex down and its base up. Along the base, the great vessels connect with the heart's chambers. Ao is the aorta; VC is the superior vena cava; RA is the right atrium; RV is the right ventricle; PA is the pulmonary artery; PV is the pulmonary vein; LA is the left atrium; and LV is the left ventricle.*

the left. These landmarks of base and apex will be used to locate other heart structures.

Previously we traced the flow of blood through the heart from the right atrium to the aorta. But this was just a flow pattern without consideration of how the blood is actually pumped through these structures. Let's trace out this pumping process.

The Cardiac Cycle

Just listening to heart sounds suggests that the heart's pumping activities could be divided into two segments: filling and contraction. But this is an overly simple view, blurred by how fast things happen and an inability to hear all that is happening. If we go through the heart's pumping cycle step by step, the heart as a pump will emerge.

Based on evidence made available through X ray and ultrasound (a means of making images of the heart with very high frequency sound), the heart's cycle can be divided beyond just filling (called *diastole*) and contraction (called *systole*). At least one more level of division is needed to keep all the events logical and connected. The mechanical heart cycle looks like this: the start of mechanical systole is atrial contraction. During this time, the ventricular inflow valves (the mitral valve and tricuspid valve) are open, and the last stage of ventricular filling is occurring for both ventricles. Shortly after the atria contract, the ventricles begin to contract and as intraventricular pressure rises, the pressure differential between the atria and the ventricles closes the inflow valves. For simplicity, let's focus on just the left ventricle for the rest of the cycle.

Pressure in the ventricle is still lower than in the aorta and the aortic valve remains closed. This period extending from closure of the mitral valve to the opening of the aortic valve is a time when the volume of the left ventricle cannot change (blood is an incompressible fluid), but the ventricle can and does change shape. This period of constant volume is called *isovolumic contraction* (*iso* meaning constant).

When the ventricular pressure finally exceeds the aortic pressure, the aortic valve opens and the heart pushes blood into the ascending aorta. This period of actually forcing blood into the aorta is called the ejection period.

At the end of the ejection period, the heart begins to relax and ventricular pressure quickly falls below aortic pressure, closing the aortic valve. The pressure in the ventricle may be below the aortic pressure, but it is still above the atrial pressure, so, once more the heart enters an isovolumic period. The atrium has been filling during these ventricular maneuvers and the pressure in the left atrium is slightly elevated. As ventricular pressure quickly falls with muscular relaxation, it eventually falls below the atrial pressure, the mitral valve opens and ventricular filling begins again. Atrial contraction soon follows, and the pumping cycle is complete. The cycle now has four primary segments in mechanical contraction: filling, isovolumic contraction, ejection, and isovolumic relaxation. These four events are shown in Figure 3–6.

Earlier, we went through the flow pattern in the heart. In comparison, clearly, flow is one thing and the mechanics of the pump are another. And it is the failure of the mechanical events in the heart that can often detour the normal flow pattern and cause disease. But some additional facts emerge from this mechanical detail.

First, the valves turn out to be passive structures that open and close simply because of pressure or kinetic energy differences on both sides of the valve. Secondly, the heart is indeed a mechanical pump that relies on valves and the physical contraction of its tissues to impart kinetic energy that finally moves blood around a vascular system.

With the heart's mechanical processes divided into four segments, a good question might be, Why could we be fooled into dividing the heart into two functions based only on the heart's sounds? A look

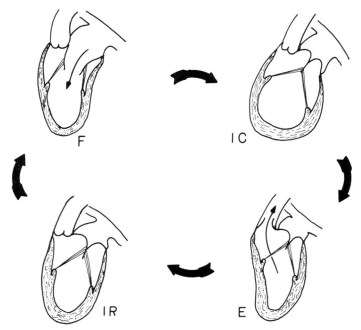

Figure 3.6. *The four basic parts of the cardiac cycle. The cycle begins with filling (F) as blood moves from the atrium to fill the ventricle. Ventricular contraction begins to build pressure within the ventricle, closing the atrial-ventricular valve. During this period (IC), pressure builds in the ventricle and the ventricle changes shape but not volume, a period called isovolumic contraction. When ventricular pressure exceeds the aortic pressure, the ventricle ejects blood into the outflow artery (E). When the pressure in the outflow artery exceeds the ventricle, the outflow valve closes, and the heart goes through a period of isovolumic relaxation (IR), leading again to filling.*

at the sources of the sounds will show how they could not show the cardiac cycle parts.

The Heart Sounds

The source and nature of the heart's sounds have been the objects of investigation and controversy for as long as humans have listened to their own hearts and the hearts of other living things. Investigation into the question has reached a rather high level of intensity over the last few years. New technologies have helped some, and although we have a good idea of the source of heart sounds, an exact mechanical explanation is still elusive. On the other hand, the timing among events is well-known.

The use of *echocardiography* (echo-ranging off the heart's structures with ultrasound in a manner similar to SONAR) provides an

THE LIVING PUMP AND ITS FLEXIBLE PIPES

unequivocal timing for the coincidence of heart sounds and mechanical events. The first heart sound is coincident with closure of the mitral and tricuspid valves. Closing these valves also marks the beginning of tension development in the ventricles. Although the mitral and tricuspid valves clearly contribute to the sound, other vibrating structures set into motion by tension development also contribute to these heart sounds [16].

The second heart sound is coincident with the closure of the aortic and pulmonic valves and some other vibrating structures [16]. Now it is clear why we could be confused. These sounds appear only with valve closure, not valve opening. The heart sounds and their coincidence with the heart events are shown in Figure 3–7.

Heart sounds are one of the primary means by which an attending physician can gain information on the condition of the heart. Along with the physical examination is usually an electrocardiogram or ECG. The ECG is made by using a set of electrodes placed on the limbs and chest to record the electrical events coming from the heart. By looking at the electrical events and how they are made in the heart, we can also gain some insight on how the heart controls its own rate.

Electrical Heart Excitation

In Chapter 1, we found out that excitable cells could move potassium and sodium to their advantage, with potassium moving into the cell and sodium out of the cell. By letting potassium slowly leak out, a cell can produce a transmembrane potential, which sets the cell interior negative with respect to its exterior. The collapse of this membrane

Figure 3.7. *The correlation between heart sounds and the cardiac cycle. The upper rectangles list each of the major events within the cardiac cycle. The heart sounds (HS) are on the lower line. The valves creating the sounds are listed at the arrows.*

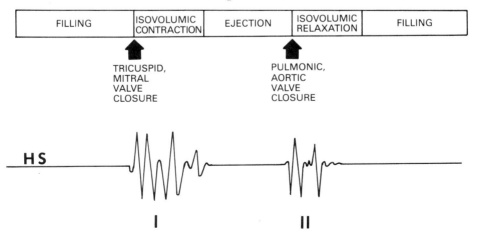

potential forms the action potential (these events are also discussed in more detail in Chapter 2). As an action potential passes along a membrane, the transmembrane potential decreases, often going several millivolts positive during the process. When the action potential sequence is complete, the cell membrane returns to its resting condition with a negative interior with respect to the outside. This same sort of action potential coordinates the mechanical contraction of the heart. And when the sequence of voltage changes are collected together over the whole heart, they form the electrocardiogram that represents a look at the electrical activities of the heart.

Normally, the heart's contraction and rhythm originate from a group of cells that spontaneously depolarize at their own rhythm. Quite properly, these cells are called *pacemaker cells,* and they regularly discharge, setting the whole heart rate. The pacing cells are grouped together in the right atrium, located near the junction between the superior vena cava and the wall of the right atrium [17]. This collection of cells is called the *sinoatrial* or *SA node.* An action potential started by the SA node spreads over the right and left atria through a set of conduction paths that finally coverge on another area of specialized tissue, called the *atrioventricular* or *AV node.* In this manner, the electrical signals that are the heart's command-to-contract communicate to the cells of the atria and eventually reach the AV node. From the AV node, the command-to-contract will be sent to the rest of the heart. The heart's major conduction pathways are shown in Figure 3–8.

At this step in the cardiac cycle, the atria are depolarized and therefore negative with respect to the rest of the heart. Because the charge over the heart is not uniform in value or polarity, an electric field quickly surrounds the heart. It is this field that the ECG electrodes sense on the limbs and the chest. This electrical event of atrial depolarization appears on the ECG as the P-wave.

The ventricles receive the depolarization as the command-to-contract through the tissue that reaches out from the AV node into the ventricular muscle. The tissue that carries out this communication is called *Purkinje tissue* and forms the Bundle of His. The tissue extends down the interventricular septum from the AV node and quickly divides into two branches that continue down the interventricular septum. The tissue then rapidly branches and penetrates into the muscle tissue of both ventricles. The branch extending to the right heart is called the right branch; the branch extending to the left heart is called the left branch. Each branch divides into increasingly finer limbs that continue to diverge into a lacework of communicating tissue that submerges into the muscle. This anatomy offers a means of rapidly communicating the action potential to the ventricular tissue, telling the heart ventricles when to contract.

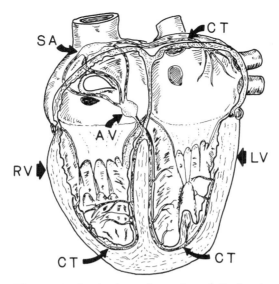

Figure 3.8. *Conduction pathways through the heart. The heart's electrical signals for contraction start at the sino-atrial node (SA) located on the right atrium, and travels over conduction tissue (CT) to the left atrium. The signals also travel to the atrio-ventricular node (AV), which connects the signals to conducting tissue (CT) called the Bundle of His, which carries the command to the right (RV) and left (LV) ventricles.*

The traveling depolarization takes time, and while the ventricles are depolarizing, the atria have repolarized, again changing the charge distribution over the heart. The charges are larger than before and opposite in polarity. On the ECG, events form a sharp spike called the *QRS complex*, composed by the sum of individual electrical waves from the heart. The QRS complex is made from three waves called the *Q, R*, and *S waves*. Together the three waves indicate that the electrical command-to-contract has reached the ventricles and spread into the heart muscle.

In order to be ready for the next contraction, the ventricles must quickly relax and repolarize. Strangely, the repolarization occurs not in the sequence of the original depolarization, but in exactly the opposite sequence. Thus, the last segments of muscle to depolarize are the first to repolarize. The result is the formation of a small positive going wave that indicates ventricular repolarization called the *T wave*. All the waves of the ECG are named right down the alphabet from P to T, which makes them easy to remember.

The events represented by these electrical waves can be summed up in the following manner: the electrical P wave signals atrial depolarization; the time delay between the P wave and the QRS complex indicates how long it took for the depolarization to travel from the SA node to the AV node; the QRS complex signifies the depolarization of the ventricles; and the T wave indicates the repolarization of the ventricles. The composite wave form is shown in Figure 3–9.

Although the electrocardiogram is an electrical signal for the whole heart to contract, the command is sent to the heart cells, which respond at the cellular level. The contractile machinery is arrayed within each heart cell, quite visible under the microscope as a set of striations within each cell. The biochemical events composing muscle contraction were discussed in Chapter 2. Just as in the skeletal muscle cell, the visible striations are the contractile proteins actin and myosin that move past one another to shorten the cell.

Although contractile proteins are the same in the heart cell as in the skeletal muscle cell, the heart cell has a general organization different from the skeletal muscle cell. For example, the nucleus of the heart cell is located near the center of the cell, not to the side as in the skeletal muscle cell. Also, heart cells lack the highly developed sarcoplasmic reticulum found in skeletal muscle. As a result, the heart cell has no well-defined storage site for calcium, which controls the molecular events of contraction. That means calcium must enter a muscle cell from its exterior with each beat. This calcium flow forms a calcium current through the cell membrane with each beat. This calcium current gives the heart's action potential an unusual form. Except for this calcium current, the remaining contractile events within the muscle cell are common to both heart and skeletal muscle.

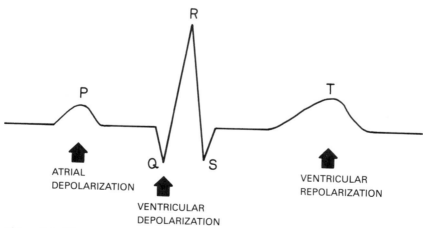

Figure 3.9. *The composite electrocardiogram. As the electrical command to contract spreads over the heart, regions of depolarization and regions of polarization present a complex, changing pattern of charges. The electric field from these charges form the electrocardiogram or ECG. The ECG waves are labeled from P through T, and the associated events are shown.*

Like skeletal muscle, the myocardial cell has an ability to change its strength of contraction as the cell length changes. The result is a length–tension curve that describes just how the muscle tension changes with muscle length. In the case of the heart, however, this curve can show the difference between a healthy and a failing heart. If the heart happens to be working on the wrong portion of this curve, the result can be a condition called congestive heart failure. And if the heart happens to be working too far along the curve, events can be virtually impossible to turn around. Let's take a closer look at why.

A length-tension curve for heart muscle is shown in Figure 3-10. The graph shows the amount of tension that can be developed by a piece of cardiac muscle as the resting length of that piece changes. Increasing the muscle length increases the tension developed by the heart muscle. This constantly increasing tension has two components: first, the active tension developed by the contractile machinery inside the muscle, and second, the passive tension developed by the connective tissue used to hold the muscle cells together. These components combine to form the total tension measured in a strip of heart muscle.

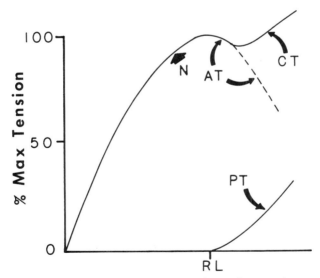

Figure 3.10. *The length-tension curve for cardiac muscle. Because the same molecular events occur in cardiac muscle contraction as skeletal muscle, the cardiac muscle has a length-tension curve much like skeletal muscle. The active tension (AT) builds and decreases, but the passive tension (PT) comes into play much sooner in cardiac muscle, and the combined tension (CT) climbs much faster than skeletal muscle. The normal operating point for cardiac muscle is not its resting length (RL), but on the ascending portion of the curve (N).*

Under normal circumstances, the heart operates on the left side of the curve as shown in Figure 3–10. As a result, when the heart volume increases at the end of filling, the myocardium responds by contracting harder, giving a more forceful ejection of blood with each beat. Should the heart be weakened by disease or injury, however, it cannot compensate for performance demands that exceed what it can deliver, and the heart volume at the end of filling increases further. Now the heart is operating on the right side of the curve peak, and as the heart volume increases, the heart's ability to eject the blood volume decreases. The result is a larger and larger heart, and a process called *decompensation*. With each increase in size comes a reduced ability to pump blood, and the heart fails, congested with blood; hence the name, congestive heart failure. With a decrease in stroke volume (the amount of blood ejected with each stroke), comes a decrease in cardiac output (a product of the stroke volume times the heart rate), and other parts of the body begin to suffer and fail as the amount of blood reaching the tissues decreases. We will later look at the drugs able to help a failing heart.

The heart is a pump that provides energy to move blood through the circulation. The heart beats and performs its functions because it is alive, composed of many living cells that can contract in a self-regulated manner. The heart has valves that normally prevent blood flow in the wrong direction and do not impede flow in the right direction. Despite the heart's central and powerful role in the circulation, it does not do it all. A second pump also helps keep blood circulating through the body.

THE SECOND PUMP

The first pump is so obvious that we might never search for a second pump. The second pump is a bit more subtle and much quieter than the first, but like the first, it has all the required elements of a pump. For example, a pump needs a muscle component to provide kinetic energy to the blood. In the second pump, the skeletal muscles provide this contractile component. The pump also needs a set of valves to keep the blood flowing in the right direction. In the second pump, the vessels provide this valve component. Let's look at how this pump is formed and operates.

The valves for the second pump are located in the larger veins that return blood from the extemities [18]. Regularly spaced along the vessel length are small cusps that rapidly open if blood should try to flow away from the heart. In fact, it is easy to show these valves with a little experiment using one of the veins located close to the surface of the skin. Applying an occluding pressure to a vein with a finger will stop the blood return to the heart in the vein. As flow stops, the back pressure will close the upstream valve nearest the occlusion, and the vein will present a distended section from the point of pressure to the valve and a collapsed section extending from the valve to the upstream portions of the vein.

Deeper veins also have these valves, and are, at the same time, exposed to the intramuscular pressures produced by contracting skeletal muscles. Contracting skeletal muscles can generate a great deal of pressure within the whole muscle. This pressure is so high that it easily closes both arteries and veins [19]. The valves in the veins, however, keep the blood flowing toward the heart, despite skeletal muscle contractions. Experiments with a device called a tilt-table demonstrate the importance of this muscle pump mechanism.

Normally the larger leg muscles work against the force of gravity, and these muscles undergo both frequent and periodic contractions. A tilt-table, however, can remove the need for leg muscle contractions and remove the second pump from the circulation. The experiment is set up by first securing an experimental subject to the table in a supine position (the patient's back is on the table). The table is then slowly tilted upright while monitoring the subject's heart rate and blood pressure.

As the tilt becomes more upright, the blood pressure begins to drop as gravity pools more and more blood into lower arteries and veins. As a portion of the central blood volume moves into the veins of the legs and lower abdomen, the baroreceptors located at the division between the internal and external carotid arteries and at the aortic arch sense a drop in blood pressure. The vascular control mechanisms that help maintain the blood pressure cause the heart rate to increase and at the same time, the arteries to narrow. As the compensation becomes more complete, the early drop in blood pressure finally returns to normal. Mild tilting along with gravitational forces brought about a vascular redistribution of blood and a subsequent decrease in blood pressure, which the system corrected back to normal. But the experiment is not yet over.

If the experimental subject is cooperative and able to keep the leg muscles relaxed over the remaining period of the experiment, the blood pressure will again fall, but more dramatically as the tilt continues. This time, however, the fall in blood pressure can lead to *syncope* or fainting.

The first drop in blood pressure is understandable, but reasons for the second drop in pressure are a little more obscure. Some rather detailed measurements show, however, that the second drop in blood pressure is due to a failure of the skeletal muscle pump. Coupling an increasing tilt with a lack of leg muscle activity, a large amount of blood pools in the leg vessels, leaving little blood to return to the heart, so little, in fact, that the subject will often faint. Because the brain normally seems to be operating on the fine edge of adequate blood flow, any sizable decrease in cardiac output can cause immediate subjective effects like spots before the eyes, dizziness, or even a faint.

Such results appear when the muscle-pump fails on a tilt-table, but are these effects visible outside the laboratory? People faint for various reasons, such as emotional stress, sickness, or side-effects from certain medications. The underlying mechanisms are the same for all three outwardly different causes. Common to all three is an insufficient blood return to the heart because of blood pooling in the lower-body vascular beds. It is not a good idea to hold up someone who is fainting. A faint is usually caused by too little blood reaching the brain because, in turn, not enough blood may be reaching the heart. Good first aid is to place a fainting victim flat with elevated legs to help restore the blood return to the heart.

Perhaps the most common place to see failure of the skeletal muscle pump is on the parade ground, where participants (both marching and watching) are required to stand at attention or simply stand quietly for long periods of time. If a person's knees are locked while standing, muscles that would normally work against gravity do not. With little movement, the force of gravity is transmitted directly through the bones and very few muscles have to be used to hold the body upright. Now add some heat that requires the body to move a large volume of blood to the skin to maintain the internal body temperature. Moving the blood volume to the skin decreases the amount of blood in the heart and great vessels. When blood flow out of the heart falls below a level required to keep the brain working, the valiant victim topples over. Experienced marchers and watchers frequently flex their legs while standing, working leg muscles. They can stay upright by effectively using the second pump to stay in position without fainting.

So far, the cardiovascular system appears to be quite dynamic, even if the owner happens to be lying down. A different sort of control takes over when the heart is stressed with physical exercise.

EXERCISE AND THE HEART

Even as children, most of us have experienced the increase in heart rate that comes with exercise. Adult joggers and exercisers are also familiar with the sensation of a fast-beating heart. A suitably fundamental question to ask is: Why should the heart speed up its rhythm during exercise?

When the body begins to work and expend energy, the energy requirements for the skeletal muscle system become quite large. Muscle cells need oxygen to make ATP molecules, and quite simply, the vascular system delivers this oxygen. Increasing oxygen delivery means that the heart must increase the amount of blood it pumps, because the maximum amount of oxygen the blood can carry per unit volume is fixed. The heart can increase its output in two ways. First, it can increase the volume of blood pumped with each beat, that is, by increasing stroke volume. Second, the heart can keep a constant stroke volume and increase its heart rate. Because the parameter the heart is trying to control is blood flow through the heart or cardiac output, and because the cardiac output is a product of stroke volume and heart rate, increasing either heart rate or stroke volume or both provides increases in cardiac output.

Elevating an exercise level increases the products of cellular metabolism including carbon dioxide in the circulating blood. The brain senses this condition using chemical sensors or *chemoreceptors* and the brain stimulates an increased respiratory rate. The segment of the brain that senses and controls respiration is the hypothalamus. But it takes time for the metabolites to reach the brain and for vascular chemoreceptors to respond. The result is a time delay between the production of carbon dioxide and its detection by the brain. Respiration is usually a little behind the body needs. The whole system of heart, vessels, and lungs, however, is now geared up to provide more oxygen to the blood (increased respiration) and more blood to the exercising tissue (increased heart rate and cardiac output).

During exercise, the distribution of blood in the arteries and veins changes, and a portion of the blood volume shifts from the venous side to the arterial side of the circulation. Vascular beds that previously had limited flow now open up, and a portion of the blood volume moves into these new volumes. As the body generates heat, the temperature-regulating system brings blood closer to the body surface to transfer the heat to the outside. Aiding this heat transfer is the secretion of sweat onto the skin surface, where evaporation cools the blood circulating just under the skin surface. That ruddy complexion thought to represent health is a result of blood shunted to the surface of the body for cooling. Clearly, the whole vascular system is very dynamic during exercise, shifting volumes of blood, opening new vascular channels, closing others, and transfering blood to the skin for cooling.

Supplying energy for blood flow for all these vascular maneuvers is the heart, composed of living cells that can respond to stress just as other cells in the body can. The heart, made largely of muscle, will grow or hypertrophy in response to the volume and pressure stresses of exercise. The heart's muscle becomes larger and stronger with exercise, and this strength or fitness is indicated by lower resting heart rates common to well-conditioned exercising people.

But all exercising does not have quite the same effect on the heart. For example, forms of exercise that tend to create prolonged muscle tension with little change in muscle length, called isometric exercises (*iso* = constant, *metric* = length) create an unusually high blood pressure during the exercise. The heart must work harder to pump blood during these periods of high stress. In response, the heart grows thicker walls [20]. A close relative to true isometric exercises is weight-lifting. Blood pressure increases with weight-lifting because the vascular beds in the lifting muscles are closed during tension, creating a very high vascular resistance. This sort of exercise appears as a pressure load on the heart. Runners and swimmers, on the other hand, experience a volume load on the heart during exercise. This load is an increased demand for blood flow without a substantial increase in blood pressure. This form of exercise also causes a larger than normal heart, but the walls are not as thickened as in the weight-lifter [21]. But in either case, it is the response of a living pump that permits this adaptive heart growth, called *hypertrophy.*

Cardiac hypertrophy is not always a sign of a healthy cardiovascular system. If a person should have a prolonged, uncontrolled episode of high blood pressure (hypertension), the heart will again grow in response to the stress [22]. But now the increase in size can lead to a decrease in performance. An acute episode of hypertension can appear without the victim knowing it, because very few outward symptoms are present. Prolonged hypertension can lead to cardiac hypertrophy. Hypertension can come from causes like renal disease or from a quite unknown cause like essential hypertension. The result is a heart that is mechanically stressed by the increased vascular resistance and that must work harder than normal to keep a proper cardiac output. This sort of hypertensive stress often leads to an altered myocardium. Just why is not clear. Some very subtle signals and responses are being sent to the heart by hypertension and exercise, signals that we do not yet understand. Just now, the observations are limited to the response of the heart to these differing growth signals.

Even before physical exercise begins, the autonomic nervous system, in response to signals from the conscious level of the brain, increases blood flow and the heart rate. This reaction is not to exercise, but to the anticipation of exercise. The strength of each heartbeat increases, and measurements of the heart's function during this time indicate that the heart appears physically stronger. This rapid change in heart function is mediated by the autonomic nervous system, and the nervous system continues this function right into the onset of exercise.

If nerves are this important to heart function, what happens in a heart transplant patient who has lost nervous influence to the heart because of the surgery? A heart transplant patient will have an increase in heart rate with exercise, but the pattern is not exactly like a normal, intact heart [23]. The question now is: If a transplant patient does not have normal neural connection to the autonomic system, how does the heart know to increase its function with exercise? The answer is: additional, non-neural indicators of exercise are brought into play.

The normal heart uses more than just nerves to signal an increased function for exercise. Nerves seem to be an early warning system, rapidly sending signals about short-term needs to the heart. For example, if vascular carbon dioxide should increase above normal during exercise, and at the same time, the oxygen level in the blood decreases, these changes in blood chemistry are detected by special chemical sensors called chemoreceptors, located close to the baroreceptors. These sensors feed information directly into the autonomic nervous system. In response, the autonomic nervous system stimulates the adrenal medulla to secrete adrenaline and noradrenaline. They, like the autonomic nervous system, signal the heart that exercise is under way and that heart rate and other vascular changes will be needed. The response of the adrenal gland is a little behind the direct nervous control, producing a short-time delay in the heart's response that appears in the transplant patient [23]. It appears that the body likes to signal events through several different communications channels, often on different timetables. These control functions will appear later when we look at the body's regulators.

Considering all the changes that exercise can bring to the cardiovascular system, is exercise bad for the heart? It seems that a certain segment of the medical community believes so [24]. A controversy is growing over whether jogging is a form of exercise that can reverse or prevent the growing number of heart attacks in this country. Despite the arguments mustered against exercise, the data seem clear. An exercise that stimulates reasonable growth of the heart increases both the heart's strength and function. But because the heart is muscle, it must be treated with the same concern and respect given to any other muscle in the body. An exercise that can overstress the skeletal muscle system and generate sore and pulled muscles is not only uncomfortable, but also potentially injurious to the muscular system. Good sense dictates that we move into such exercise schemes slowly, giving muscles time to grow in strength and capability. The heart as a muscle should be treated no differently. A heart that is out of condition should not be pushed into heavy service too quickly.

The heart needs time to respond to the long-term demands attendant to exercise. Muscle growth produces more muscle tissue, and the heart needs more blood vessels to supply these larger cells with nutrients and oxygen. An inadequate circulation to an exercised skeletal muscle will cause it to fatigue early. The heart struggles with the same problem. A slowly increasing exercise program will give muscle tissue time to develop adequate circulation in both skeletal and heart muscles. In short, any exercise program should begin slowly and cautiously, giving the heart and other tissues time to respond and grow. And growth takes time because the heart must partition its available energy into the physical contraction on each beat, as well as any additional demands of growth, and only a finite amount of time and energy is available.

Some hearts, however, have been damaged by disease or developmental defects. Let's look at some of the problems of a diseased heart for the physiology they illustrate.

THE DISEASED HEART

Heart disease in one form or another strikes a very large number of Americans yearly. When it happens, the pump, so vital to life, decreases its function, sometimes a hardly measurable amount, other times completely. If the decreased function is large enough, the victim becomes a cardiac invalid. The period of recovery can be short if only a small portion of the heart muscle is involved, or it may be quite long if a large part is destroyed. And heart disease does not strike an inanimate object; rather, it strikes a living tissue by denying that tissue fuel and oxygen needed to stay alive. Fortunately, along with our understanding of heart physiology comes a realization that alternatives exist that permit a victim of heart disease to resume some of life's normal activities. For others, too much of the heart is lost, and they have to alter life in a major way to suit the limited function of the heart.

Difficulties can begin right at the start of the heart's cycle: the action potential that provides the command-to-contract. The signal to contract begins at the SA node and travels to the AV node. The conduction path to the AV and the AV node itself can be disrupted by disease. If the electrical pattern of conduction over the heart is distorted, portions of the heart can be electrically separated from one another, and each portion will begin to function independently. The heart may then take up a rhythm too slow or too irregular to maintain adequate blood supply to the brain. The heart has lost its normal pacemaking capabilities. If a very low heart rate is the problem, a new pacemaker must be placed in the heart, a pacemaker that is man-made. This new pacer is electrical, running off self-contained batteries. Some pacers can take on rather high levels of sophistication, not only setting the heart's rhythm, but also increasing the heart's rate on demand.

The idea behind implanting a pacemaker is rather simple: when the normal mechanisms of pacing the heart fail, a new artificial pacer is placed in the body to control the heart. Many people are now alive and working because of artificial pacemakers. But because the pacers are electronic, they have a sensitivity to certain forms of electromagnetic radiation, which prompts all the warning signs around public microwave ovens. Whether pacing is natural or man-made, the heart must beat with a rate high enough to supply blood to the brain and the body.

Although the heart is made largely out of muscle, parts of the heart very sensitive to disease are the heart valves. Unlike muscle, the valves are made of connective tissue, and are quite thin, considering the stress they receive with each beat. Bacteria can sometimes attack valve tissues, leaving a disfigured and malfunctioning valve. The scars on the valve after an infection can make normally free leaflets fuse together, producing a *stenotic valve.* If the valve is very stenotic, the inability to move blood properly changes the blood distribution in the heart's chambers as blood pools upstream of the stenosis. For example, a stenotic mitral valve will produce a dilated left atrium [25]. And the diseased valve brings disease not only to the heart, but to the whole body, as reduced blood flow impairs the ability to support the whole body.

Valves not only need to open properly, but they must also close properly. A bacterial infection can leave the boundaries of the valve leaflets scored by scar tissue, preventing the leaflets from closing properly, compromising the valve's water-tight seal. Bacteria can even erode a hole through a leaflet. With either condition, the valve becomes leaky, and the net cardiac output decreases as blood flows back through the defective valve. This sort of valve dysfunction often produces a sound or murmur, which the attending physician can hear with a stethoscope.

A valve that fails to close properly is called *incompetent.* Like stenosis, an incompetent valve will cause blood to pool in heart chambers and eventually distort the normal anatomy of the heart [26]. But whether or not the valve happens to be stenotic or incompetent, if it causes systemic distress, endangers the life of the patient and cannot be repaired with surgery, the only recourse is to replace the valve with a functioning artificial or prosthetic valve.

Prosthetic valves are now a real success story in medicine. More are being used and through this experience, the clinical community is learning how to have fewer complications. An incompetent or stenotic valve can have pronounced effects on the heart and body, and many of these effects vanish with valve replacement. A new, prosthetic valve can restore an adequate blood supply to the whole body. The patient resumes a more normal body weight, and can often carry out mild exercise. Without a knowledge of materials and how they interact with blood, all this would be impossible, for the real danger in introducing foreign materials into the blood is the formation of blood clots. The right materials for the prosthetic valve have made valve replacement a success.

Although heart valves are fragile and subject to disease, the most frequent form of heart disease is the heart attack [27]. This common name really describes two forms of disease. First, a heart attack can be caused by a coronary artery occlusion formed by an *embolus*, usually a portion of a blood clot that has traveled to the coronary circulation, resulting in a complete loss of blood flow to a portion of the heart muscle. Second, a heart attack can be caused by the formation of *atheromas*, which are small deposits of cells and fatty material that narrow the vessel diameter and decrease the amount of blood that can flow through the artery. These two forms of disease are similar in that a reduced or interrupted blood flow is common to both, but the causes are not identical.

The heart is composed of cells that need oxygen to live. When the blood supply is denied the cells, they die, and that portion of the heart that is dead is called an *infarction*. If the infarction is large, the heart may not be able to compensate for the loss of contractile tissue, and the heart will fail. If the heart survives the initial loss of tissue, the dead cells are eventually replaced with stiff, connective scar tissue. Although this new tissue cannot take an active role in the heart's contraction, it does provide support and strength to the heart wall while the remaining muscle contracts and pumps blood. As a result, the maximum amount of work the heart can carry out may be reduced, but a careful exercise program can begin to restore some of the heart's capacity.

Soon after a heart attack, when the infarction has healed with scar tissue, many cardiologists begin the patient on a heart-building exercise program. The intent is to strengthen the remaining portions of the heart to restore part of the heart's functional capacity. Through a program of diet and exercise, the attending physician attempts to reduce the possibility of another clot formation and an ensuing second heart attack in the patient. Whether or not such a program will really permit the patient to live longer is unknown, but restoring a heart attack victim to a more normal way of life carries its own reward.

The second form of heart attack produces an intense pain called *angina pectoris*. The pain comes from a heart muscle starved for oxygen *(hypoxia)*. The reduced tissue oxygen level results from a narrowing of a coronary artery, reducing blood flow (called *ischemia*) to a portion of the heart's muscle. Insufficient blood flow prevents the compromised portion of the heart from keeping up with its functional demands.

A reduced blood flow is often a consequence of *atherosclerosis* of the coronary arteries. The deposits not only narrow the vessel lumen, but also prevent the vessel from going through normal adaptive changes in diameter. The vessel is hardened, which prompts the more popular name for this condition, "hardening of the arteries." These fatty deposits frequently build up in preferred locations such as vascular bifurcations and sharp vascular bends [28]. As a result, the coronary arteries have a decreased blood-carrying capacity.

The cause of atherosclerosis is not currently known. It appears to be associated with several conditions, among them hypertension, smoking, diet, and family history, but no clear single cause or combination of causes has been found [29].

Several new theories seem to explain many of the observations. One of the latest is the so-called tumor theory. In this model, a single smooth muscle cell becomes mutated and grows like a tumor, narrowing the vessel diameter [30]. Local cellular activity increases the production and aggregation of cholesterol and fatty acids. Recent attempts to determine the origin of atheroma cells show that the cells all have the same parent cell. If this model is correct, then chemical compounds in our environment that are *mutagens* (compounds that can cause mutations in growing cells) or carry mutagens like cigarette smoke could contribute in a major way to the formation of atherosclerosis. One source of mutagenesis is viruses that are able to make some cells abnormal [31]. Whatever the cause or causes, atherosclerosis as a pathology is found in a widening number of people at increasingly younger ages [32].

Once formed, an atheroma does not seem to regress, even if the patient should change life styles. The only way of effectively dealing with an atheroma currently is to remove it surgically, direct blood flow around it, or mechanically expand the stenosis. Surgical removal of an atheroma is not now a practical solution to coronary heart disease. Redirecting the blood flow around the blockage and expanding the stenosis are, however.

Redirecting flow around a blockage is called a bypass. In this operation, a section from another vessel is sewn into position to shunt the circulation around the stenosis. For a coronary artery, the technique is called a coronary bypass. The result is often an effective restoration of circulation to heart muscle, and the angina attacks that haunted the patient before the operation either disappear or are markedly reduced. But an operation will only change the physical course of blood, not the underlying causes of disease, and often a new bypass vessel also experiences atheroma formation and narrowing [33]. Heart transplant patients have similar problems with atheromas building up in the coronary vessels of their new hearts. And the present evidence points to an inability to change the course of events with drugs, diet, or physical therapy. There is much yet to learn about the causes of atherosclerosis and how to cope with the disease once it is in progress.

But the heart and vascular system is one of the most responsive systems to treatment with drugs of various kinds. Some of these agents and their effects will shed some light on the physiology.

CARDIOVASCULAR PHARMACOLOGY

Drugs that work on the heart draw a lot of interest not only because so many people use them, but also because the drugs usually have understandable effects. A drug may make the heart beat faster, slower, more regularly, or just more strongly. Taking a specific substance at regular intervals can calm an excited heart, strengthen a weakened heart, or reduce a dangerously high blood pressure.

One of the most widespread disorders of the cardiovascular system is *hypertension* or high blood pressure. Heading the list of causes is "essential hypertension," which is really a name rather than a description. The cause of essential hypertension is unknown. This form of hypertension begins with an increased activity of the autonomic nervous system, which narrows the arteries, causing the blood pressure to rise. The narrower vessels cause an increased workload on the heart, which is pumping into these narrowed vessels. From this increased workload, the heart begins to grow. As the pressure increases even more, small capillaries that cannot withstand the pressure begin to burst in organs and tissues like kidneys, the retina of the eye, and the lungs. Weakened vessels become tortuous and distended with pressure, and can burst or balloon out, forming *aneurysms.* Hypertension takes a toll on the heart and on the rest of the circulation.

If vessels do not break or balloon out, the victim can suffer headaches, dizziness, and a high probability of forming blood clots, which lead, in turn, to heart attacks and cerebral vascular accidents. Studies have shown that people with low or moderate hypertension do not, on the average, live as long as people with normal blood pressure [34]. These findings suggest that even moderate hypertension must be treated to prevent the more dangerous complications of stroke and heart attack.

Fortunately, the organization of the autonomic nervous system and the smooth muscle in the vessels provides several opportunities to effectively intervene and decrease a rising blood pressure. Figure 3–11 shows the basic organization of the autonomic neurons and the vascular smooth muscle. Each site in the signal flow that uses a neurotransmitter to transmit information is a potential site for drug intervention. The most convenient is the last synapse between the nerve

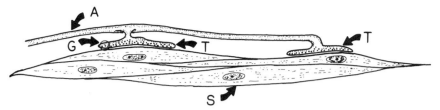

Figure 3.11. *The synapses between autonomic nerve terminals and vascular smooth muscle. The vascular command to contract travels along autonomic nerve fibers (A) to terminals (T) that form synapses with smooth muscle cells (S). The neurotransmitter is stored in granules (G) found in the nerve terminals. This synapse turns out to be a good site for chemical intervention to treat hypertension.*

terminal and vascular smooth muscle. At the nerve terminal are several chances to intercept an elevated nervous activity.

Electronmicrographs of nerve terminals show small clusters of dark granules inside the nerve that decrease in number when the nerve is stimulated into activity. The neurotransmitter, in this case norepinephrine, initiates smooth muscle cell contraction and modulates the amount of contraction. The muscle cell receives this message when norepinephrine molecules bind to receptor sites located on the surface of the muscle cell. This molecular binding is a reversible chemical reaction, so, after a norepinephrine molecule occupies a receptor site for a period of time, the reaction can reverse, releasing the neurotransmitter, leaving it free to move about. Several alternate fates are possible for these free molecules.

First, they could diffuse away from the synapse and eventually enter the general circulation, where enzymes in the blood plasma inactivate the neurotransmitter by breaking the molecule into two or more pieces. Very often, however, the molecules diffuse back toward the nerve terminal, which is physically very close. At the nerve terminal, the nerve cell moves these molecules back into the nerve using molecular pumps located in the cell membrane. Recovering the norepinephrine lets the nerve save energy that would otherwise be required to make new neurotransmitter molecules. This re-uptake of the neurotransmitter by the sympathetic nerves provides a means of affecting the functioning nerve terminal.

This cycle of events at the nerve terminal offers three ways of slowing down a hyperactive nerve.

First, we could send in a molecule that looks like norepinephrine, but does not work when it binds to the muscle cell receptors. This molecule would block the neurotransmitter right at the smooth muscle membrane.

Alternatively, we could flood the nerve terminal with molecules that prevent it from taking up the free neurotransmitter molecule. Because re-uptake of free neurotransmitter molecules is a major contributor to the total content of the storage site, reducing the amount of stored neurotransmitter reduces the amount of released neurotransmitter, which, in turn, reduces smooth muscle activity.

On the other hand, we could send in a molecule that looks like the neurotransmitter molecule and that can be taken up like the transmitter molecule, but that, once it reaches the storage site, hangs on longer than normal. This strategy again reduces the number of neurotransmitter molecules available for release by an active nerve terminal.

Common to all three methods are molecules that have a shape close to the real one, and therefore look like the transmitter molecule. The look-alike does not act like the transmitter at the various receptor sites, however, altering the connection between nerve and muscle in the bargain.

Although the autonomic nerve terminal is an attractive place for drug intervention, not everyone responds well to this treatment. In this case, other methods of reducing blood pressure can be used either in conjunction with these neuroleptic drugs (drugs able to change the activity of neurons) or in some cases as an alternative to the drugs. For example, reducing the blood volume in the vascular compartment can reduce blood pressure. Blood volume is reduced by decreasing the amount of water in the vascular compartment. This can be brought about by stimulating the kidney to excrete more sodium and water than usual. Agents that stimulate the kidney to excrete more water and sodium are called *diuretics*. For many, it is the combination of a neuroleptic and a diuretic that effectively controls blood pressure.

A drug that interferes with the normal function of the sympathetic nervous system provides additional insight into how the overall regulation of the cardiovascular system works. For instance, a drug that intercepts the operation of the sympathetic nervous system depresses the cardiovascular system even more than if the whole nerve trunk to the heart were cut [35]. Experiments show that for any given state of the circulation, two antagonistic systems, the *sympathetic* and *parasympathetic systems*, regulate the heart, and no one of them is ever completely silent. A continuous state of operating antagonism exists between them. The final state of the circulation, then, is defined by the balance between the two. This sort of antagonistic yet balanced control can be found in other systems in the body, and is described in more detail in Chapter 4, on regulators of the body.

Because the body uses two antagonistic nervous systems to determine the final circulatory state, operating extremes can go in either direction. Thus, just as some people can have hypertension, others can have the opposite of hypertension, hypotension. For these people, the vascular system is unable to compensate for changes in posture, and they cannot rapidly stand from a sitting or lying position without fainting. Considering the pharmacology up to this point, perhaps the most logical approach to this condition would be to introduce a drug to the vascular system that acts and looks like norepinephrine. In practice, this sort of therapy seems to work, but for only a short period of time [36].

Prolonged therapy relies more on setting up a strategy for a hypotensive to use when getting up from a chair or bed. In addition, garments that compress the lower body and legs work well. These garments are patterned after the anti-gravity suits used by jet pilots to prevent blackouts during high-speed maneuvers in fighter aircraft. The suit increases lower body pressure and prevents blood from being distributed away from the heart and central circulation. Compression suits are a mechanical yet effective way of solving the problems of blood distribution in hypotension.

Blood distribution problems can arise when the heart begins to pump inadequately. A heart in failure is a heart dilated and congested with blood. The heart's ability to contract is severely reduced, and without some help, the heart will fail. As it turns out, the heart can often be brought back to a near-normal function, despite the underlying disease, with the use of some unusual drugs. They are classed under the heading *cardiac glycosides*. The most common is *digitalis*.

Digitalis is a drug that the physician must prescribe with care. In small doses, it is a great healing agent, bringing a sick, failing heart back toward normal. Adding more digitalis, however, makes it function as a myocardial poison, bringing about heart arrhythmias (irregularities in the normal heart rhythm) and an abnormal heart function. The margin of safety between effective doses and systemic poisoning is relatively narrow, requiring caution for both physician and patient.

Digitalis comes from several different sources. One of the most common is the foxglove plant *(Digitalis purpurea)*, and one cardiac glycoside is even found in the skin glands of a toad. In the long history of foxglove use in humans, it has entered periods of acceptance and rejection, mostly as a result of the narrow range of safety for the drug, a range not well understood by early physicians. But current understanding makes it a very effective means of treating a failing heart.

Like many other agents, digitalis also provides indications of how the heart works. In appropriate doses, digitalis seems to work on the myocardium (the heart muscle) directly, increasing the heart's strength of contractions [37]. The heart not only beats more strongly in response to the drug, but also increases the rate of tension development. Just how digitalis is able to cause such changes in the heart is not yet understood.

Along with these muscular changes, digitalis seems to affect several components of heart cell function. ATP-dependent pumps in the myocardial cell membrane, for example, are decreased in activity, leading to less sodium being pumped out of the cell and less potassium being pumped into the cell. The result is a reduced separation between the threshold for an action potential and the membrane resting potential. Does this help the heart? It is clear that moving the resting membrane potential closer to the threshold does increase the potential for spontaneous activities in the heart, which can lead to heart arrhythmias, a condition the physician wants to avoid. But digitalis also causes the calcium flux into the heart to increase. It may be that separately or in concert, these cell changes produce the whole heart's response.

Strangely, the beneficial effects of digitalis appear only in a weakened heart. Supplying digitalis to a normal heart produces no effects at normal therapeutic doses [37]. Despite our ignorance of just how digitalis carries off its biochemical effects, it is a very effective tool in contemporary cardiac therapy.

Let's consider for a moment what an attending physician might see when treating a failing heart with digitalis, with an eye on the physiology it reveals.

Because the heart is failing, blood pools in the venous circulation, creating a high capillary pressure that forces fluids from the capillaries into the tissues. If the right heart is failing, the patient expresses a condition called *dropsy*, which is edema (a movement of fluids into the tissue spaces outside the vascular system) in the dependent portions of the body. If the left heart is failing, blood first pools in the lungs, causing a condition called *pulmonary edema.*

The first effect of digitalis is to restore the balance between the right and the left sides of the heart [38]. Tissue edema decreases, either reducing the body size or relieving the labored breathing of pulmonary edema. The patient looks better, feels better, and can often become more active.

Heart failure is not the only way for a heart to lose its ability to pump blood. An irregular heartbeat can also cause a loss of blood flow to the vascular system. The sources of an irregularity are quite varied. Some heart arrhythmias can be dangerous, leading to complete disorganization of the heartbeat, setting the stage for a lethal event. Other irregularities, however, may be part of the heart's sensitivity to a particular drug, or to a mild form of heart disease, or to an innocuous set of heart events. For the most part, these irregularities are inconvenient, but not life-threatening.

A rather common form of arrhythmia is a *sinus arrhythmia*. In this pattern, the heart rate changes as a person breathes. The changes are a normal response of the heart, adapting to the amount of blood returning to the heart. Inhaling decreases the pressure within the chest, and venous return to the heart increases. The increased amount of blood stretches the SA node, increasing the heart rate. In contrast, exhaling increases the pressure within the chest, and venous return to the heart decreases, decreasing the heart rate in turn. Most dogs have a very dynamic version of sinus arrhythmia. A few minutes of listening to a dog's heart offers a good example of this sort of rhythm.

A change in the amount of blood entering the heart is not the only way to change the heart's rhythm. If the balance of ions inside and outside the heart's cells is altered, the myocardial cell may begin to generate action potentials at a very high rate, increasing the rate of the whole heart. This elevated heart rate is called *tachycardia*. If the pacing cell for this tachycardia is located in one of the atria or the AV node, the condition is called *supraventricular* tachycardia. If the pacing cell is depolarizing too fast, the heart may be unable to follow. A high heart rate may be high enough to reduce ventricular filling because of too little time between beats. A reduced filling can reduce the cardiac output, causing the patient to faint when the blood supply to the brain falls below a maintenance level. The requirement is clear; the abnormal pacing cell must return to a normal rhythm. Doing this requires stabilizing the cell membrane.

A local anesthetic can stabilize a heart cell's membrane potential. At the molecular level, the anesthetic molecules block the heart cell's sodium channels that open for sodium conduction through the membrane. And sodium, as we learned earlier, is part of the early depolarization of the action potential. In addition, the fact that a local anesthetic can control a cell's permeability to sodium suggests that sodium permeability may be a major factor in the cell's instability in the first place. This controller of irregular hearts is called *lidocaine*, a close molecular relative to the novocaine used as a local anesthetic in dentistry [39].

Lidocaine works so well over such a broad range of dosages without being toxic that it can be used intravenously to help treat the more dangerous heart arrhythmias. It is also valuable in reversing the arrhymthmias produced by digitalis intoxicaion.

Irregularities in heart rhythm can also arise when a region of the heart receives insufficient blood flow, producing a region of ischemia. The ischemia produces pain along with an irregular heartbeat. The ischemia condition becomes very evident when exercise or excitement causes the heart to beat more vigorously, increasing its own perfusion (the amount of blood flow used to maintain tissue function) needs. To reduce the pain, either blood flow to the ischemic area must be restored, or the heart's workload must be reduced. A rapid solution is to open the systemic arteries as quickly as possible, reducing the heart's workload. Because vascular smooth muscle can be influenced through the vascular system, getting something into the blood that will relax the vascular smooth muscle is a very direct approach. The compound that does all this is *nitroglycerine*. Nitrates like nitroglycerine dilate systemic vessels containing smooth muscle by causing the muscle cells to relax. From this molecular activity we get another glimpse of vascular control and the relationship between arterial diameters and the workload on the heart.

Understanding the heart as a pump requires understanding the load the heart must work against and how it senses this load. The cardiovascular system as a whole is trying to keep both regional and systemic flow adequate. Despite the need to control blood flow, the vascular system does not have a natural flow sensor. As a result, flow must be sensed by following a contributing factor to blood flow, in particular, blood pressure. Spotted throughout the circulation are neurosensors picking up regional pressure or volume changes in large and small arteries [40]. Combining pressure with a vascular diameter defines the blood flow through a vessel. Our earlier equation for flow indicated that if a vessel diameter increases, a lower pressure can keep the same amount of blood flow. A lower pressure means a lower workload for the heart. And reducing the heart's workload reduces the pain of angina pectoris.

CONCLUSIONS

The heart is an audible organ, generating sounds with each beat. Alive and self-regulating, the heart adapts to changing body conditions to keep blood flow adequate for tissue needs. The heart pumps a tissue called blood through a set of flexible pipes called arteries and veins. The pipes are not just passive structures; they contribute actively to the blood flow that reaches capillaries deep in vascular beds.

Disease in the system can cause discomfort, debilitation, or death. To respond to these threats, we have not only the heart's own recuperative powers, but also a collection of drugs that can help reverse some symptoms, aid the heart in beating more strongly, or take the load off a heart unable to keep up with its work demands.

We can strengthen the heart with exercise, treat it with drugs when it is sick, and even replace it with a new heart under the right conditions. It beats each day of our lives, giving life and seeking little rest. It may not be the seat of the soul, but it is indeed the center of our physical well-being.

4

REGULATORS OF THE BODY: HORMONES AND NERVES

The power word for life is control. For life to exist in simple, single-cell organisms in a pond, or in whales that live in the sea, and certainly in man, the forces of life cannot be random. They must be controlled and channeled. The complexity of living things can arise and be maintained only through an organization and partition of energy at the cellular level. Without control, a cell's life could be threatened if essential energy were diverted into nonproductive activities. Controls within a living organism keep the inside of its cells focused toward life activities, the outside of the cell bathed in the correct materials, and the cell clearly informed on what to do and when to do it.

Control or regulation within the human body can be divided into two major categories: hormones and nerves. Although these two regulators are far apart in some organizational aspects, they only infrequently function without one another. And in some instances, special organs stand as intermediaries that transform nervous activities into hormonal activities, becoming *neurohumoral transducers*. These organs bridge the separation between nerves and hormones, bringing them together to work in concert.

Of the two, hormones seem to have more draw on human curiosity, despite the extensive work in neurobiology. The nature of hormones has been vigorously pursued by the scientific community, which has discovered a growing complexity of activities unpredicted by early ideas about hormones. Hormones are fascinating because they make things happen using the power of only a few molecules, which may circulate in the vascular system for long periods of time or flicker into existence for only a few moments and vanish. Hormones tell the body to grow, tell it when to take on adult form, tell it to turn up the cellular metabolism to maintain a core temperature, or even set out the time for reproduction. Hormones provide a biological shield for the many stresses that shape human life today. They are, indeed, powerful compounds.

Nerves, on the other hand, have activities that are more visible to a researcher. Nerves are living cells, quite specialized in activity. They do not move within the body and generally cannot reproduce themselves when they reach a final functional form. They live, it seems, only to communicate. They do so in the brain on a grand scale. But without controls, a nerve cell could not work in a world of communications noise generated by itself and neighboring cells. Nerves need internal controls while exercising external controls on other cells. In general, nerves bring a type of control that is fast and "tight," yet carries a great range of flexibility.

It seems a paradox that the freedom of life requires rigid controls on one hand and regulation on the other. And in this regulation, the left hand does need to know what the right hand is doing. Body-confusion and disease can arise if key functional information is wrong, lost, or misdirected. People with brains that are surgically separated into two noncommunicating hemispheres (used to defeat a very severe form of epilepsy) show some of the qualities of confusion we are concerned with. These people react as if they had two brains, because the two hemispheres of the brain operate without the benefit of knowing and communicating with each other [1].

On a more practical basis, other forms of confusion less evident than a separated brain can be seen when scientists have outward signs that they can detect or measure. To understand how to use outward signs to detect a hormone that may be present in quantities too large or too small to detect directly, or to detect a nerve that begins to act independently of its functioning unit, we have to spend a little time looking at control systems.

Control systems, unfortunately, can become quite complex. For our purposes, however, they will be treated in simple terms. Before this chapter is finished, we will come to use many control systems concepts as tools to understand the way the body brings together small cellular acts that form the mosaic of activities we call life.

We looked at nerves earlier in Chapter 1 on cells, and again in Chapter 2 on the skeletal muscle system. Hormones have not had the same exposure. So, before looking at control systems, let's get an overview of hormones.

MOLECULES AS MESSAGES

The number of cells within a human or other large mammal is astronomical. The average human has about 10^{14} cells at any one time. For all these cells to act together, they must communicate with one another. Cells communicate through the language of molecules, and here the medium is the message. If a message molecule travels some distance from its sending or secreting cell, the molecule qualifies as a hormone [2]. Hormones are widely present in all animals. Comparative physiologists have found only a few examples of cells that communicate primarily through electrical events, despite an excellent potential to do so. The bulk of cellular communication seems to be carried out by the transmission and reception of molecules that are messages.

But how far should a molecule have to travel to be a hormone? And how long should it exist to be a hormone? Hidden beneath the electrical events of nerve and muscle are communicating molecules that carry messages over a distance so small that calling the molecule a hormone is logically difficult. We call it a *neurotransmitter* as a result. Ignoring the distances involved, neurotransmitters have many of the properties imparted to hormones. In general, a communications molecule is a hormone only when the distances traversed are large, and is called something else when the distances are very small [2].

At the outset, the idea of using message-carrying molecules may seem like a poor choice as a communicating medium. A little analysis, however, will reaffirm the wisdom of this choice.

First, the number of different molecules that could be made by a cell is very large; therefore, the number of molecular messages available to the cell is equally large. Second, each sort of molecule can have a unique shape, making the messages equally unique. Through the variety and specificity of the molecules, a message uniqueness can extend to both transmitting and receiving cells. Third, the distance between a transmitting and a receiving cell can range from a few millionths of a meter to several meters. A molecular message is not likely to get garbled just because the message happens to be traveling a long distance. As much cannot be said of electrical communications.

So, hormones are messages passed among cells, and neurotransmitters are also messages passed among cells, but spanning a very short separation between cells. Regardless of distance, however, each molecule conveys a message in the language of cells.

The messages carried by a hormone are predetermined. The molecule has a meaning accepted by both transmitting and receiving cells. Random communication is no more acceptable at the cellular level than at the whole organism level. The patterns of this controlling communications system are the next subject in this exploration of life regulation.

REGULATING CONTROL SYSTEMS

Control systems can take on a wide range of forms. One of the simplest is the "regulator," a device that works to hold a variable close to a set value. It is an organization used widely in the body.

Communication is a large part of regulation. Parts of the regulator must work together to carry on the process of regulation. In like manner, hormones, both as messages and as parts of a control pattern, tell portions of an organism what to do and when to do it. From the pattern comes not only the statics of organ regulation, but also the dynamics of organ adaptability. The term "static" in this sense is used to denote an ability to hold a body parameter such as body water within the normal limits essential to life, despite pressures from inside and outside the body that try to force body functions away from normal. Dynamics, on the other hand, will denote body activities like movements, contractions, filtrations, and secretions. Whether dynamic or static, the control systems have a very similar form.

Although some control systems can take on high levels of complexity, many others are simple and show the essential relationships we will need in order to examine more complex patterns. A good example of such a simple system is the toilet tank and the way it fills to a fixed water level after the tank is drained. The goal is to have enough water in the tank to flush the bowl when needed, but not so much water that the tank would overfill. As water drains out of a tank, a large internal float falls with the water level, opening a valve controlling water flow into the tank. Through the filling valve, water enters the tank, replacing that which drained away. As the tank fills, the float follows the rise in water level, decreasing the filling rate at the same time, until the correct water level finally closes the filling valve. The ring around the inside of the tank at the standing water level attests to the accuracy of the system, despite its simplicity.

The relationship between the valve and water level is termed *negative feedback;* that is, as the water level increases, the control pattern causes the influx of water to decrease. So, the control system in a toilet tank is an example of regulation and negative feedback that is both simple and accurate.

With this simple relationship in mind, let's look at a more complicated system called a negative feedback controller. It has a few more elements, but still holds many of the original toilet tank relationships. The organization is shown in Figure 4–1.

The analysis begins at the far right of Figure 4–1 at the *controlled variable.* A controlled variable might be a substance in the blood or a mechanical output like the thickness of a board. Regardless of the variable, the idea is to control that variable to a narrow set of values that center on an ideal value.

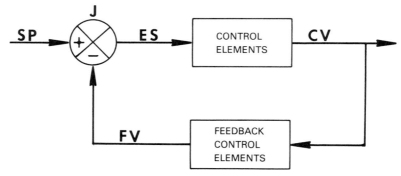

Figure 4.1. *The schematic feedback controller. The goal of the system is to control a variable (CV) around a value determined by the set point (SP). The existing values of CV are fed back through the feedback control elements to form the feedback variable (FV). This variable is compared with the set point in the summing junction (J). The difference between the two forms the error signal (ES), which drives the control elements.*

In turn, the *control elements* produce the variable, changing the amount of the variable in response to an input command. If the control elements should produce too little of the controlled variable, the input command would increase production. If the variable becomes too much, the input command tells the control elements to decrease production. In Figure 4–1, the input command that increases or decreases the amount of the controlled variable is the *error signal.*

Assume for a moment that the bottom half of Figure 4–1 did not exist. The input command that controls the production rate of the controlled variable would be the *set point.* Increasing the set point would increase production; decreasing the set point would decrease production. But the question arises, How does the set point "know" that the control elements are responding correctly to the input commands? The only way it can do that is to look at or sample the controlled variable to see if it is at the right level.

The system must sample the condition of the controlled variable and bring or feed it back to the "set point." Usually, however, the energy form of the controlled variable and the input command (set point in this case) are not the same. Clearly, then, the system must sample the controlled variable and change it into a form that can be compared with the set point command. The *feedback control elements* make this conversion, producing the *feedback variable.*

Comparing the feedback variable and the set point with one another occurs at the circle in Figure 4–1, called a *summing junction.* By comparing these two signals, the error signal does not have to be a continuous signal. Comparing the set point and the feedback variable by subtracting one from the other will produce a zero error signal when things are right, and either a positive or negative error signal when things are wrong. Two pieces of information reside in the error signal now: the direction to change the controlled variable and how much of a change is needed.

Up to now, it has been a rather abstract system. It is time to look at how this controller regulates. Consider first what happens if an outside influence makes the controlled variable increase above its operating level. The controlled variable value passes through the feedback elements to become the feedback variable, with a value above the set point level. The error signal becomes less than zero, that is, it becomes negative. The positive or negative sign on the error signal tells the system which way to move. The amount of error signal, on the other hand, tells the system how far to move. As the controlled variable tends toward normal, the error signal moves toward zero.

On the other hand, if the controlled variable becomes lower than normal, the signal feeding back to the summing junction decreases, producing a positive error signal. The positive error signal tells the control elements to increase the controlled variable value. This increases the control feedback signal and the error again returns to zero. Thus, sampling the controlled variable and comparing its value against a reference sets the whole regulator to work.

With two models in hand (the toilet tank and the controller), we can approach some of the biological systems with a means of identifying elements and understanding the control process. Our first physiological system will be the apparently uncomplicated sugar–insulin system.

THE SUGAR-INSULIN SYSTEM

The sugar–insulin system is a good starting point because it has a feedback path as direct as the toilet tank. This system controls the blood sugar through a hormone called *insulin*. The primary sugar handled by the system is glucose, but other sugars such as sucrose are also managed by this system.

Insulin is a hormone secreted by pancreatic cells that respond to the amount of sugar in the blood. Because glucose is a primary fuel for metabolism, the object is to regulate the blood sugar around a point adequate for normal body function. Figure 4–2 shows how the various segments are interrelated.

The pancreatic cells release insulin when sugar molecules bind to the cells' membranes at sites that can receive only sugar-shaped molecules. These receptors permit the cell to sense both the presence and the amount of sugar in the blood. A rule of proportions governs the event. The cell functionally estimates the amount of sugar present in the blood by responding in proportion to the number of occupied

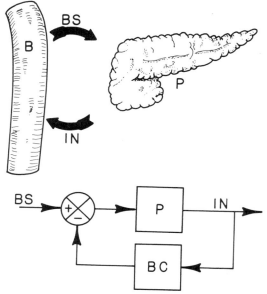

Figure 4.2. *The blood sugar-insulin system. The pancreas (P) secretes insulin (IN) in response to the blood sugar level (BS). Insulin, on the other hand, decreases BS, which in turn, decreases the secretion of insulin. This forms a one-sided controller, which has no way of handling external decreases in blood sugar. BC are the body's cells.*

receptor sites at any one time. As the concentration of sugar in the blood increases, the amount of sugar bathing the pancreatic cells increases and so does the number of occupied receptor sites. As a consequence, the amount of secreted insulin follows the concentration of blood sugar.

This system represents a control with features much like the toilet tank. The events are quite parallel. For example, as blood sugar and the tank water level increase, the insulin and inflow valve slowly bring things back to a resting state. In theory, the parallel between the toilet tank model and the body sugar–insulin system permits an investigation of the regulation properties of the biological controller (the sugar–insulin system) by looking at the mechanical controller (the toilet tank).

But this look is only preliminary. Unlike the toilet tank model, the insulin system has a few complications that can hide some of the functional elements.

The sugar–insulin system works on a simple proportion: more sugar, more insulin; less sugar, less insulin. The regulation point for sugar management is not zero, just as the toilet tank water level is not zero. Instead, blood sugar is managed to a "normal" level that supplies enough glucose for the brain and other organs to continue working.

The brain, in contrast to other tissues, is not influenced by insulin. Regardless of how much sugar is available or how much insulin, the brain tries to extract a constant amount of sugar from the bloodstream, which it can do if the blood sugar level is normal or above normal. If available sugar is lower than normal, the limit on the extracting process is the prevailing blood sugar level. Whether resting, asleep, or thinking very hard, the brain's rate of sugar extraction appears to be constant [3]. If the blood sugar level drops below normal, however, the brain is soon not working to capacity. Clearly, brain function relies directly on how wisely the body manages its blood sugar level.

Although this system seems rather comfortable and simple, a little analysis shows that it is really too simple. No real set point control seems evident in the regulator shown in Figure 4–1. Further, the action–reaction pattern for insulin described up to now is not really conducive to good control. The toilet tank has a similar problem. It has no way of handling too much water. To regulate, biological systems need more than a one-sided control. Living systems often handle this requirement by stimulating an antagonist to the primary control at the same time. The insulin system has such an antagonist, a hormone called *glucagon*.

Glucagon produces an effect opposite of insulin, that is, it increases the amount of blood sugar. A new organization including this new antagonist is shown in Figure 4–3. Using the second hormone, glucagon, the body is able to respond not only to a blood sugar level above normal, but also to a blood sugar level below normal.

In the complete system of insulin and glucagon, it is the balance between the two hormones that determines the net amount of glucose finally entering the cells.

This sort of antagonism is frequently found in biological settings where the pattern of a true regulator does not exist. But along with antagonist and agonist, the blood must receive a supply of sugar, for if no new sugar were to enter the blood, reducing the movement of sugar into the cells would not be enough to raise the blood sugar level. The whole process is like the ancient Chinese concept of Yin and Yang, two natural but antagonistic forces in the body that establish the balance of life through their balance. Imbalance produces disease. That is certainly a good description of diabetes.

Diabetes

No single word can really describe the diseases called diabetes. In general, diabetes is a condition in which the pancreatic cells have a reduced ability to make insulin, or have lost the ability entirely. Under

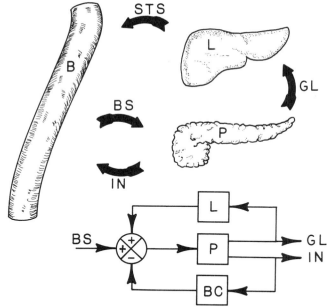

Figure 4.3. *The expanded sugar-insulin system. The final system does not look like a true feedback controller. Instead, two antagonistic control paths exist with a set point input for blood sugar (BS). The antagonistic hormone is glucagon (GL), operating through the liver (L) to release stored sugar (STS). P is the pancreas; IN is insulin; BC are the body's cells.*

the general heading of diabetes are two separate and distinct forms of this disease. One is called *juvenile onset diabetes;* the other is called *adult onset diabetes.* The two names describe a common condition but different times for disease onset. The different times also suggest different underlying mechanisms at work for each condition.

Juvenile onset diabetes is characterized by a complete inability to produce insulin in any amount. Without a replacement for this lost insulin, death is inevitable. In contrast, adult onset diabetes is often less severe, although it can advance to a situation that requires insulin replacement. Usually, the adult form of diabetes is controllable by a diet that regulates the amount of sugar entering the body and drugs that help the secretion of insulin from unproductive pancreatic cells. And as the division suggests, research indicates that these two different forms of diabetes may have two different causes.

An explicit cause for juvenile onset diabetes is presently unknown, but some current studies show a pattern in the detection of diabetes which suggests an infective agent, perhaps a virus. Juvenile onset diabetes is growing faster than the general population, and the number of new cases discovered each year follows a distinct cyclic pattern, presenting periods of highs and lows for new diabetics [4]. This pattern looks like another one for a well-known viral infection. If the cause is a virus, it may operate on the pancreatic cell the same way a polio virus operates to kill nerve cells [see Chapter 2 on skeletal muscles]. If a single infective organism is involved, it might be possible to use an immunization program to decrease the incidence of this disease. Regardless of the cause, a newly discovered juvenile onset diabetic must live with insulin replacement for life.

Adult onset diabetes, on the other hand, appears to have a different cause. Evidence points to a strong genetic influence on the incidence of this disease [5]. For example, it tends to follow family lines. Progression of the disease could almost be described as a "tired cell" syndrome, as the pancreatic cells gradually lose the ability to make and secrete insulin.

A direct form of therapy, then, is to stimulate the pancreatic cells into additional secretion or match the blood sugar to the capabilities of the pancreatic cells. Severe forms of adult onset diabetes do occur, and require more than simple stimulation; they require total replacement of insulin.

On one side, diabetes is a very real disease, but on the other side, the disease provides some clues about how the body obtains energy from sugar and fat, and in some measure, the proportion of energy from each.

Three energy substrates are available for the body: sugars (carbohydrates), fats, and proteins. At the same time, the different body tissues have different dependencies on each of these substrates. For example, the brain uses only glucose, whereas the skeletal muscles use all three energy substrates all the time. The heart, at the same time, obtains about 80 percent of its energy from fat, with the remainder coming from glucose and proteins, with proteins making up a very small percentage [6]. An untreated diabetic is unable to get sugar into the insulin-sensitive cells, which forces the cells to rely more on fat and proteins for energy. As the cells use more fat, they produce a large residue of compounds called *ketones* [7]. Ketones appear in the urine and are excreted through the lungs. An untreated diabetic's breath takes on the odor of acetone as the ketones build up in the body. But an acetone breath is only one of several clues to what is going on.

Carefully watching the course of diabetes indicates that insulin replacement alone is not enough. Even with careful management, the diabetic suffers greater than normal incidence of heart disease, vascular disease, and blindness [8]. We evidently understand only part of this condition. What happens when the antagonism of glucagon is lost? And might more hormones be involved? The questions currently outnumber the answers.

The sugar–insulin system does not have a biological summing junction, which is part of the basic controller model. Most of the other hormone systems do. So, before looking at some of these more sophisticated systems, let's gather some facts about the biological summing junction.

The Biological Summing Junction

Essential to the theoretical function of a controller is some sort of summing junction or comparator. Considering the function of the comparator, a biological analog to this device might have little chance to exist. That notion may explain why the control systems theory had slow going in biology. Control theory said it ought to be there; biologists called it a search for a mathematical construct. But it was not just theory. The biological summing junction exists, and a look at some of the more sophisticated control systems will show how biologists missed it for so long.

The biological summing junction was very likely first discovered in neurobiology. It turned out to be a neuron. In the brain, neurons are connected in great arrays of convergent and divergent patterns. These patterns of connection among neurons permit the physical relationships needed in a summing junction. It turns out that a convergence of several neurons into a single neuron can produce both additive and subtractive processes [9]. For example, the influence of one neuron on another can be either excitatory (additive) or inhibitive (subtractive). It follows that if both excitatory and inhibitory inputs arrived at the same neuron at the same time, that neuron's function would be the effective sum of both inputs.

An excitatory input is one in which an increase in activity of neuron A produces an increase in activity of neuron B. The arrangement for this connection is shown in Figure 4–4. On the other hand, an inhibitory process is one in which an increase in activity in neuron A produces a decrease in neuron B. Figure 4–4 shows this sort of arrangement, too. With relationships such as these in neurons, it is no surprise to find neurons in nearly every regulator in the body.

The discovery of the biological summing junction represents more than just the search for a confirmation of theory. It represents a change in the tools of thought. The entry of the hard sciences such as physics, chemistry, and engineering into biology was difficult. The reason can be found in the principal language used in the hard sciences: mathematics. To a largely observational science, mathematics had no real value outside simple statistics. The need for mathematics was inevitable, however, because the living cell lives in a physical world, governed by the same laws of physics and chemistry that can be observed

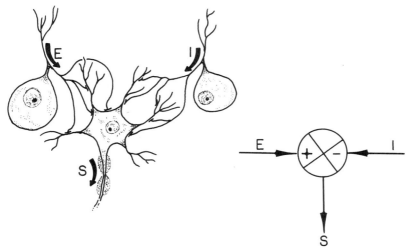

Figure 4.4. *The biological summing junction. Early workers on biological control systems felt that the summing junction was only a mathematical construct. Neurons, however, can form summing junctions with excitatory (E) and inhibitory (I) inputs to a common neuron, forming a summary output (S). The neurons are on the left; the effective result is on the right.*

and measured outside the cell. Physiology turns out to be, in part, physics in a living system.

Fortunately, mathematics is not needed to understand how a biological regulating system gains power through organization. We can look at organization and appreciate the interconnecting patterns among elements. Looking at control systems in the body, however, will eventually bring us to a "super organ," a small pink appendage at the base of the brain, which turns out to be the center of control for nearly all glands in the body.

THE MASTER GLAND

The title for this section is really not very original. On the other hand, it is difficult to select another title because this one, although a cliché, is still the most accurate. This small gland, surrounded by bone and brain, simply has power far beyond what its size would suggest. The gland's power comes from hormones, which reach out to other tissues and glands throughout the body.

This gland is called the *pituitary* and is positioned at the base of the brain. The location is strategic, placing the pituitary at the interface between the brain and the rest of the body. The gland functions as a neurohumoral transducer, changing neurological information into hormonal information.

The neurological information begins within the brain, and through a specialized vascular pattern, the pituitary sends hormonal messages to the rest of the body. With only a few exceptions, pituitary hormones generally affect other glands, which in turn transmit other hormones to the body. It is an elaborate but effective pattern, ensuring an accurate transfer of hormonal information to the body.

Despite a well-known anatomy, the significance of the pituitary's organization escaped translation until modern methods for measuring very small quantities of hormones entered the scene. A schematic of the gland's anatomy is shown in Figure 4–5. Central to the transducer process is a vascular pattern that joins the function of two tissues that make up the organ.

The gland consists of two segments designated the *anterior* and *posterior pituitary*. Histology shows these segments to be different tissues with different origins [10]. The anterior portion comes from non-neural tissue, whereas the posterior portion comes from neural tissue. The vascular bed that fingers through these two tissues provides a unidirectional communications system between the two. The vessels form two capillary beds that invest each portion of the gland. The

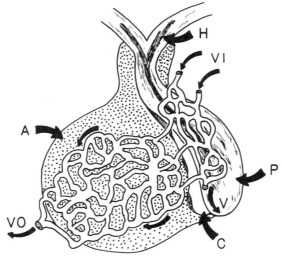

Figure 4.5. *A schematic view of the pituitary gland. The pituitary functions as a neurohumoral transducer, changing neurological information into hormonal information. The posterior pituitary (P) has connections with the hypothalamus (H). The posterior pituitary releases "hormonal releasing factors" that enter the first circulation. The releasing cause hormone release from the anterior pituitary (A) into the outgoing vessels (VO). Some direct secretions from the posterior pituitary can reach the circulation directly through V, bypassing the anterior portion. C is the pituitary cleft.*

vascular beds, in turn, are in series. A serial capillary system such as this is called a *portal system*. It is through this vascular system that the pituitary receives its instructions from the brain, and in turn communicates these instructions to the rest of the body.

Despite an unusual anatomy and molecular communications, we have a good picture of how things occur within this gland. One system of which we have an especially good understanding is the pituitary–thyroid system. Before bringing the whole system together, however, a starting point is to first look at the various components of the system. Understanding the components will make it easier later to bring all the components together into a single system. The thyroid is the beginning point.

The Pituitary–Thyroid System

The thyroid gland sits in the front of the neck, saddling the trachea and the esophagus. It is normally not very large, and has a pink appearance from a profuse vascularization that surrounds and penetrates through the tissue. The gland secretes two hormones that have a common molecular origin and shapes that are nearly alike. In fact, they differ only by the attachment of a single iodine molecule.

The hormones begin with the amino acid *tyrosine*, and acquire three or four iodine atoms along the synthesis pathway. The compound with three iodines is called *tri-iodothyronin* or T3. The compound with four iodines is called *tetra-iodothyronin* or T4. To make these molecules, the thyroid gland gathers and stores iodine from the blood. If the thyroid should fail to get enough iodine, it responds by growing large to improve secretion of T3 and T4.

The two *thyronins*, T3 and T4, regulate the body's metabolism by controlling biochemical events within the mitochondria [11]. As a result, a thyroid over- or under-producing the thyronins will affect overall body function. For instance, an increase in T3 and T4 in the body turns up the body's metabolism, heating up the body, and producing additional side effects along the way [12]. At the mitochondrial level, the thyronin decouples the production of ATP from the electron transport chain, expending the cellular energy into heat rather than ATP. As a result, food energy that would normally go to maintenance of body weight is diverted into this accelerated activity. The result is an individual with a low body weight who is overly sensitive to external heat. In this manner, hormones T3 and T4 alter body activities by directly influencing the cellular chemistry.

Controlling these thyroid hormones is another hormone secreted by the pituitary gland. This hormone stimulates the thyroid gland secretion, so the hormone is called *thyroid stimulating hormone* or

TSH. Within the molecular shape of TSH are several cellular instructions that include: trapping more iodine; synthesis of more thyronin; and the release of more thyronin. The two hormones, TSH and thyronin, are coupled to work together; that is, as TSH levels increase, thyronin levels increase as well. Clearly, then, if the secretion of TSH should become out of control, the whole body would feel this imbalance.

TSH, like thyronin, is also controlled, but from two different inputs: first from the circulating levels of T3 and T4; and second from another hormone that travels only a short distance. The patterns of feedback for this control system are starting to take form.

The short-range hormone has such a short distance to travel that it has another name. It is called a releasing factor. The releasing factor comes from nerves extending down from the hypothalamus into the posterior pituitary. These nerves secrete the hormone (releasing factor) into the first capillary network of the organ's portal circulation [13]. The releasing factor combines with cells in the anterior pituitary by traveling through the second capillary system. The releasing factor stimulates the release of TSH. The full name of this neurotransmitter is *thyroid stimulating hormone releasing factor*. That name is accurate but a bit unwieldy to use all the time, so the hormone is known as TSH-RF instead. With a releasing factor in one hand and the remaining components of the system in the other, the whole system can come together.

The brain sets the amount of releasing factor that can enter the first capillary network in the portal circulation. Any releasing factor in the portal blood finds receptors in the anterior pituitary tissue, through the second vascular bed. The releasing factor stimulates TSH secretion into the systemic circulation. TSH reaches the thyroid through the systemic circulation and stimulates both the production and secretion of thryronin. Thyronin, in turn, reaches the body's cells through the circulation, and at the same time reaches the brain and pituitary gland through their vascular systems. The arrival of thyronin in the pituitary and brain causes a decrease in the amount of TSH released. The complete feedback relationship is shown in Figure 4–6.

The organization shown in Figure 4–6 holds all the basic elements of the model controller first shown in Figure 4–1. The brain sets the operating level; the pituitary transduces the brain's instructions into hormonal statements that tell the body to function. The circulation closes the loop, transporting TSH and thyronin for control and effect.

When this system fails or moves away from normal, the results were first thought to be a sign of beauty, then a sign of disease. The beauty sign was a smooth, bulging throat, which we now call a *hypertrophied* (enlarged) thyroid gland [14]. Still, if the gland became too large, it was no longer a source of beauty. An enlarged thyroid is a *goiter*, which has walked a fine line between beauty and disease for

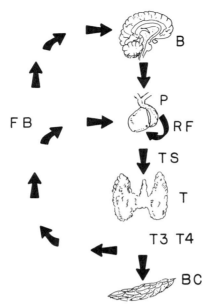

Figure 4.6. *The pituitary-thyroid gland control pattern. Primary instructions for thyroid secretions start in the brain (B). They reach the thyroid gland (T) through the pituitary gland (P), which secretes a releasing factor (RF), causing the release of thyroid stimulating hormone (TS). The thyroid secretes T3 and T4, which feed back (FB) on the anterior pituitary and the brain.*

many years. Too little iodine in the diet sets the growth process in motion. A look at the thyroid model (Figure 4–6) will show why.

The first step in making a goiter is a lack of dietary iodine. Without iodine, the thyroid cannot make T3 or T4, which leaves the gland with very little thyroxin to release into the circulation, despite the body's demand for it. A low thyroxin level in the blood signals the blood and the pituitary to secrete more control hormones, TSH and TSH-RF. With more TSH reaching the thyroid, it grows more tissue to make more T3 and T4. The thyroid becomes large trying to increase the amount of tissue that can pull iodine out of the circulation, like a ship unfurling more sails to gather more wind. If the iodine deficiency is only moderate, the compensating growth of the thyroid will give the sensuous lines of the neck portrayed by so many early European painters. In severe cases, the gland can grow to grotesque proportions. But a thyroid gland is only one indicator of an abnormal TSH secretion.

The fat pads behind the eye are as sensitive to the presence of TSH in the blood as the thyroid gland, and they grow in parallel with the thyroid. As the thyroid grows, it distends the normal smooth neck lines. As the ocular fat pads grow, the eyes begin to protrude, producing a characteristic stare associated with a goiter. Once the tissue growth occurs, it does not revert to normal size easily. Even with a corrected diet including iodine, neither thyroid nor ocular fat pads return to normal size. The thyroid often requires some surgery. In the end, a thyroid deficiency may eventually leave a scar on the neck from thyroid surgery and an aggressive stare. Or it can leave a woman with a sensuous neck line and a slightly penetrating look.

A dietary deficiency is not the only way to produce a goiter. Iodine therapy does not always stop or reverse a developing goiter. In such a case, something else may interfere with the normal thyroid controls. The sources of such interference can be often traced to biochemical agents that impede one of the biochemical steps in thyroid control. The compounds are collectively called *goitrogenic compounds.*

Compounds that cause goiters can push things out of kilter at several key points in the thyroid–pituitary system. For example, the ability to absorb iodine through the gastrointenstinal wall can be affected or the ability to trap iodine in the thyroid, or even the ability to secrete thyronin after being synthesized. Many of these compounds can be found in the foods we eat. Plants of the genus *Brassica*, which include soybeans, brussels sprouts, and cabbages, contain one such goitrogenic compound [14]. These compounds normally appear in such small amounts in the diet that they present no real hazard. But if a dietary chain amplifies the amount a person receives or if the diet becomes overly dependent on only one or a few of these plants, goiters can be a result. These goiters, furthermore, do not respond to iodine therapy, illustrating a cause other than an iodine deficiency. In the end, the goitrogenic compound must then be hunted down and removed from the diet.

Hormonal systems can take on even more sophistication than the pituitary–thyroid system. Systems can overlap, creating side effects that begin in one system and appear as an imbalance in another. Without knowledge of these connections, the results can confuse what might seem to be a simple diagnosis. One such overlapping of systems is the pituitary–adrenal cortex system and the kidneys.

The Pituitary–Adrenal System

The organization of the pituitary–adrenal system is a lot like the thyroid control sysem. Both use a similar pattern of hormone interactions that provide a sensitive and intricate control of the final hormone level. Both use a negative feedback loop that communicates through the bloodsteam. Unlike the thyroid system, however, the adrenal system extends an unanticipated finger of hormonal influence into kidney function. A general layout of the organization is shown in Figure 4–7. Similarities with the thyroid control system are obvious. Once again, the hypothalamus defines the set point for the control system. The hypothalamus uses a releasing factor to stimulate a release of *adrenalcorticotrophic hormone* (ACTH) [15], which in turn stimulates the synthesis and release of *cortisol* and another hormone, *aldosterone*. Aldosterone, it turns out, influences the retention of sodium and water in the kidneys [16]. The feedback path to the pituitary and hypothalamus is through the vascular system and is negative; that is, it inhibits the secretion of both releasing factor and ACTH. Except for the somewhat disconnected secretion of aldosterone, the overall pattern looks quite like the thyroid system.

With a sense of all the organization in this system, let's take a look at the hormones from the adrenal cortex. They are the hormones under final control and have several influences on the body.

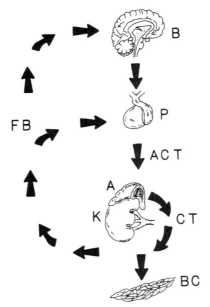

Figure 4.7. *The pituitary-adrenal gland control pattern. The primary instructions for adrenal secretion start in the brain (B) and pass through the pituitary gland (P) to the adrenal gland (A) via adrenocoritcotrophic hormone (ACT), The adrenal gland secretes cortisol (CT), which affects the body cells (BC). CT also feeds back (FB) through the vascular system to the pituitary and the brain.*

All hormones secreted by the adrenal cortex are collectively called *steroid hormones* because they all contain a common array of molecules called a *steroid nucleus*, based on the cholesterol molecule [17]. An extract of the adrenal cortex contains about 50 different steroid compounds, of which only eight seem to have biological activity.

All the hormones are formed by removing or adding small molecular side groups that extend out from the cholesterol molecule like spokes from a hub. The activity of each molecule is determined by both the location and the type of molecular side group hung onto the cholesterol molecule. For example, only one side group separates the molecules *progesterone* and *androgen*, which have widely different activities. Specifically, progesterone inhibits female ovulation and androgen stimulates male secondary sexual characteristics and the sexual drive in both sexes.

Supporting the synthesis of all these complex molecules is an elaborate array of enzymes and biochemical reactions. Despite the complexity of its biochemistry, the adrenal cortex secretes only one hormone in substantial amounts, *cortisol* [18]. Cortisol has widespread influences on the body, and the absence of cortisol is life-threatening. In general, the hormones secreted by the adrenal cortex support the body's ability to respond to stress. Without them, even a minor stress could be lethal. For example, the cardiovascular system begins to fail in the absence of cortisol. In addition, without cortisol, bacterial infections are poorly repelled by the body, permitting profound and rapid bacterial infections.

Among the corticoids secreted by the adrenal cortex is aldosterone, a hormone that influences kidney function. Some of the controls on kidney function will be examined later in detail. For now, our interest is only that aldosterone influences kidney function. Through the renal system, aldosterone stimulates the production of another hormone called *angiotensin* [19]. Angiotensin increases blood pressure and also stimulates the secretion of aldosterone. That is an odd arrangement, with two hormones stimulating the secretion of one another in a circular pattern. Fortunately, this circular interaction is an incomplete portion of the total kidney control system. Other parts of renal regulation will bring some order to the interactions of aldosterone and angiotensin.

Angiotensin is part of another sort of control scheme—one that does not use the central nervous system to set hormonal levels. Up to now the major input for control systems came from the brain. The set point for the whole system came from the brain. Physical functions like blood pressure can also set a hormonal control system into full swing. One such system is the renin–angiotensin system.

THE RENIN–ANGIOTENSIN SYSTEM

The renin–angiotensin system is a control system with widely distributed parts that helps to maintain systemic blood pressure, but through an indirect means. Systemic blood pressure, which is the entity under control, is sensed by this system within the kidney [20]. Despite the fact that the brain is excluded from this control system, the renin–angiotensin system remains one of the most powerful in the body.

Starting the renin–angiotensin system into action is a reduced blood pressure inside the renal artery. Nestled next to the renal artery within the renal capsule are a group of cells different from other renal cells. These special cells contain granules like those found in other cells known to secrete hormones and similar substances. These granulated cells are sensitive to the blood pressure within the renal artery and respond to a reduced renal artery pressure by secreting a substance into the arterial blood. Investigators found this substance to be a protein, but it did not appear to function like a hormone [21]. The protein turned out to be an enzyme that has no systemic effects of its own. Instead, it acted on another protein normally found in the blood. This protein is called *angiotensinogen*.

The hormone that actually affects systemic blood pressure is formed in a two-stage process that begins with the secreted enzyme, called *renin*, biting off a ten-amino acid sequence from angiotensinogen [21]. This ten-amino acid sequence (called a *decapeptide*) is fairly innocuous and itself carries no major influences on blood pressure. But it does carry the potential to be one of the most potent influences on the cardiovascular system ever found in the body.

The first decapeptide is called *angiotensin I*. Angiotensin I travels through the venous side of the renal circulation, finally reaching the heart. The heart pumps the hormone into the pulmonary circulation and in this circulation things being to happen. Found only in the lung tissue is an enzyme that removes two more of the ten amino acids, producing an *octapeptide* (eight amino acids), which is the final product [21]. The molecule is called *angiotension II*.

Angiotensin II comes out of the lung and goes through the heart to reach the remaining circulation. Here the molecule causes arterial smooth muscle to contract, increasing systemic blood pressure. The rise in systemic blood pressure restores the blood pressure to the renal artery and the special cells, and the cells lower renin secretion.

The degree of control and the power of this system offers a clue to the priorities the body sets between systemic blood pressure and renal blood flow. Despite the dangers of high blood pressure, renal blood flow has preference over the central blood pressure. If disease makes this system lose its balance, the result can be a very serious and prolonged high blood pressure attack. The organization of the renin–angiotensin system is shown in Figure 4–8. For completeness, Figure 4–8 includes the aldosterone response.

The renin–angiotensin system controls renal blood flow by manipulating the systemic blood pressure. The question arises: What does aldosterone secretion have to do with blood pressure? Part of the answer to that question can be found in Chapter 3 on the heart. Blood pressure is not only a function of arterial narrowness, but also the blood volume relative to the volume capacity of the vascular system. Because the holding capacity is finite, an increase in blood volume can increase blood pressure. Aldosterone tells the kidney to hold back some of the sodium that would normally be lost through urine formation. Along with sodium retention, the kidney retains water, which effectively increases the blood volume and thereby the blood pressure. In this manner, the hormone angiotensin II directly increases the blood pressure in the short term by narrowing vessels, and in the long term by stimulating secretion of aldosterone, which expands the blood volume, increasing blood pressure.

The renin–angiotensin tandem provides both short-term and long-term control of the cardiovascular system, without the brain

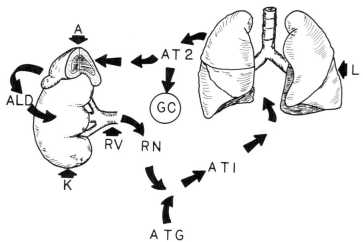

Figure 4.8. *The renin-angiotensin-aldosterone control pattern. Sometimes one hormone can cause the secretion of another to provide both a short term and long term response. A reduced blood pressure to the kidney (K) causes the release of Renin (RN) into the renal vein (RV), which interacts with angiotensinogen (ATG) to form angiotensin I (ATI). ATI goes through the lung circulation (L) and becomes angiotensin II (AT2), which affects the general circulation (GC), increasing short term blood pressure. AT2 also stimulates the adrenal cortex (A) to secrete aldosterone (ALD), which causes sodium retention, leading to a long term effect on the blood pressure.*

somewhere in the loop. For that reason alone, this hormonal system stands in contrast to other control systems in the body.

Along with the renin–angiotensin system are other mechanisms that influence circulation. For example, the autonomic nervous system sets arterial diameter and consequently blood pressure. Under the right conditions, however, the renin–angiotensin system can become dominant. But malfunctions of the autonomic nervous system can also override all other controls. The result is essential hypertension.

HORMONES AT THE CELLULAR LEVEL

Despite the varied influences of identifiable hormones, just what happens at the cellular level is as yet only partly known. Just how does the cell translate the message coded in the molecular shape of a hormone? Some portions of the system are known, but many of the remaining events are shadowy and defy attempts to know them better.

Hormonal influences start at the cellular level, often involving identifiable cellular components. Proteins that function as hormone receptors protrude through the cell wall, ready to respond when a hormone combines with the receptor [22]. One such protein receptor is an enzyme called *adenyl cyclase*. This enzyme makes a substance called *cyclic adenosine monophosphate* or cAMP from ATP [23]. This simple yet important molecule functions in the cell as a second messenger that transmits the instructions carried by the first hormone into the cell's interior. Although cAMP is the communications link, just what parts of the cell receive its message is still vague.

Some hormones, however, do not need a second messenger. They can penetrate quickly into the cell's interior and combine directly with nuclear DNA [24]. As a consequence, they bring about physical changes in the DNA structure that can be seen with the electron microscope. These conformational changes in DNA may represent the turning on or turning off of DNA segments, and thereby a sifting out of cellular activities. Very simply, controlling the cell's DNA controls the cell.

Despite the power of hormones to control body functions, other control systems use just nerve and muscle. Let's take a look at a few of these systems.

NERVE AND MUSCLE CONTROL SYSTEMS

Up to now, nerves have been central elements to a bio-control system. They were summing junctions or transducing elements or set point sources for the whole system. Nerves are so involved in so many aspects of the body that it is nearly impossible to avoid them in any discussion on control systems. But the title of this chapter suggests an examination of nerves and hormones as controllers; symmetry would dictate a consideration of control systems made only of nerves. Outside the brain, however, it seems that few if any examples of pure nerve controllers exist. Instead, nerves outside the brain are either effectors that have specific target organs which they influence, or sensors that send messages back to the brain. In the brain the target organs are other nerves. But if our rules of symmetry are allowed to slide a little and we consider nerves with organs, some interesting control systems become available for inspection.

Nearly every organ and tissue in the body is under some sort of neural control. It is evident that the involvement of nerves with so many tissues is a consequence of the primary neural function: communications.

Nerves are used by the brain to link itself with the body and the outside world. Sensors bring in information that the brain needs to control the whole living organism, and motor nerves bring out the information the brain sends to make the many control systems work. Clearly, the neuron is the prime communicating cell of the body.

To provide a broad view of nerve–muscle relationships, we will look at two nerve–muscle controllers: the motor nerve–skeletal muscle system; and neuro-cardiovascular regulation. But before getting into the organization of the discrete neuromuscular controllers, we need to look at the overall organization of the nervous system. The organization will channel some of our thinking along specific lines as common relationships among the systems emerge.

Dividing the Nervous System

On a simple anatomical basis, the whole nervous system naturally divides into two segments: first, the somatic system; and second, the autonomic system. Historically, classifying parts of the nervous system was first anatomical. When the science of pharmacology came of age, the division turned out also to be chemical. At the same time, physiologists divided along the same lines for functional reasons. It is clear the division is real despite several different points of view.

In general, the somatic nervous system deals with activities over which we have some voluntary control. An obvious example of this activity is skeletal muscle movement. The neurons that stimulate activity in skeletal muscle are called motor nerves. They arise out of the spinal cord with a particular anatomy and are called *anterior horn cells.* When these cells are destroyed by disease, the result is paralysis. And without motor nerve activity, the muscle cells atrophy, becoming weak and small.

In contrast, the autonomic nervous system deals with automatic activities of the body. This rule has undergone recent modification with experience in "teaching" control to the autonomic system [25]. We will learn more about the automatic qualities of this system in the later parts of this discussion.

Despite any disclaimers, the functions innervated by the autonomic nervous system are largely unconscious. An example of such a system is the control of blood pressure; another is the adaptation of the iris of the eye to the ambient light level and the emotional state of a person. Examples of autonomic innervation abound. Voiding, excreting, breathing, and secretions from certain glands are additional examples of functions controlled by this system. The control of blood pressure is one of the best-understood examples and will be a good choice to look at in detail later.

Neuro Input and Output

Considering either the control of skeletal muscles or of the muscles that line the vascular walls shows that they share a common organization. The control system has an input, which is a sensory nerve, and an output, which is a motor nerve.

Sensory nerves are information gatherers, working as transducers, converting energy into a form the nervous system can use. The information currency is the nerve action potential. The sensory nerve responds to a stimulus by generating a series of action potentials with a frequency in proportion to the intensity of the stimulus. Pressure, motion, heat, cold, or the presence of specific molecules (scent and taste) are transformed by sensory nerves into a stream of action potentials.

Each major form of energy sensed by the body has a special neuron to detect the presence of the energy [26]. Heat sensors, for example, sense a presence of heat; cold sensors sense a lack of heat. Nerve endings in the eye (rods and cones) respond to both light intensity and light color. In contrast, touch receptors in the arm respond very little to heat or cold or light. And following the division of input energies, each sensory nerve has a specific or proper stimulus.

From all the sensory nerves comes a massive amount of information about what is happening inside and outside the body. "The five senses" is a popular phrase, but an inspection of the body reveals many more than that.

One way to appreciate some of these additional senses is to realize just how much a body "knows." For example, with eyes closed, we still know where each limb is located with respect to the rest of the body. We can reach an itch in the dark because of this sensory input that gives limb position. These data come to the brain from sensors distributed in joints, muscles, and tendons, each funneling pieces of the total awareness about limb position to the brain.

Because data flow into the brain comes at such a high rate and in such volume, some of it is channeled away from our immediate awareness. For example, contemplating the pressure applied to the buttocks and legs while sitting or the pressure to the back while lying down can bring discomfort. Stopping to think about it probably produced some movement and reposturing by the reader. It did in the writer.

If we got up during this little awareness experiment, then the second portion of the nervous system came into play, the motor nerves. The motor nerves motivate things to happen. They reach the tissue with brain or spinal cord instructions that tell a tissue what to do. The result may be a contraction, a relaxation, a secretion, or an increase in metabolism. Motor nerves complement a set of nervous functions, sensory and motor, that make up an action–reaction sequence called the *reflex arc*. It looks simple from the outside, but looks are deceiving in this case.

Components of the Reflex Arc

Within the brain and spinal cord are an astronomical number of cells, most of them speaking to one another in patterns of convergence and divergence. Thus, we may give our brain a simple stimulus from a sensory nerve and receive a motor response in return, but the processes within the brain and spinal cord are quite complex. Some of these paths, however, are limited to only a few neurons. Others are untraceable in a maze of cells with unmapped connections.

But the reflex arc is a primary piece of knowledge lodged firmly in the physiology of the nervous system. It is so firmly entrenched, in fact, that elaborate psychological theories use it to explain all behavior. We speak of responding to someone else or something else. Humans respond to the environment, to social status, to rage or fear of another, to love from another. B. F. Skinner saw patterns in this behavior and suggested that all human activities were reflexive [27]. Certainly psychological experiments seem to suggest that many human activities, both conscious and unconscious, are reflexes in response to specific stimuli.

Still, reflex arcs are part of regulating systems. So let's look at a few of our neuro-regulators to gain some insight on how they keep us going. One of the systems contributing a great deal to our daily life is neuro-control of skeletal muscle tension.

SKELETAL MUSCLE TENSION CONTROL

The neuro-motor system for the skeletal muscle is pretty straightforward: neurons emanating from the spinal cord innervate the muscle fibers. A skeletal muscle sensor, however, has a more complicated anatomy and physiology. Let's start with the sensor.

Skeletal Muscle Sensor

The position and general organization of a sensor is shown in Figure 4–9 in schematic form. The complex is called the muscle spindle and is made of sensory nerve endings and some very special motor nerves.

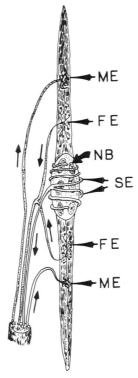

Figure 4.9. *Organization of the muscle spindle. The muscle spindle provides information to the central nervous system about tension in the skeletal muscles. The spindle is set in parallel to the muscle fibers. The system has two sensory endings, the flower spray ending (FE) and the annulospiral ending (SE), which winds around the nuclear bag (NB). The tension on the sensor is controlled by motor nerves that terminate on the motor end plates (ME) and set the tone of the muscular portion of the spindle.*

The sensory fibers terminate on the center of the structure in two locations with two distinct forms of nerve endings. One nerve ending resembles a flower spray and is called *flower spray ending* for that reason. The other sensory nerve terminal splits into a net of fine fibers that grip the cell wall in a spiral fashion. This organization generated the name of *annulospiral ending*. Any tension that changes the shape of the cell body in the nuclear bag region will either increase or decrease the stretch on the annular nerve terminal, permitting the nerve terminal to sense the amount of tension applied to the nuclear bag. At the same time, the flower spray ending senses changes in tension along the cell membrane. With these two nerve endings, the amount of stretch or tension imposed on the whole apparatus can be transformed into a set of action potentials.

On either side of the nerve sensory endings are motor nerve endings that activate the contracting cell elements and thereby change the amount of tension on the spindle. They are called *gamma efferent fibers* because of their size (they fall into a class of nerve sizes called gamma) and their motor (efferent) function. The motor nerves terminate on the cell with a normal appearing motor end plate. The motor nerves serve to keep the length of the spindle apparatus the same as that of the muscle fibers surrounding the sensor.

The operational organization for the complete system of spindle apparatus and muscle fibers is shown in Figure 4–10.

The spindle apparatus is parallel to the major muscle fibers. As a result, if the muscle is stretched so is the spindle, and the sensory output action potentials increase in frequency. This is shown in Figure 4–10 A.

Decreasing the length of the muscle from its stretched position by muscle contraction takes the stretch off the spindle apparatus, and the output from the sensory fibers goes to zero. Now the gamma efferent fibers increase their firing rate, shortening the spindle to restore the sensory discharge. This process is shown in Figures 4–10 B and C. The problem is one of following muscle length and still keeping the sensor working to detect changes in muscle activity.

If the muscle is once more stretched from its new length, the sensory output increases greatly, even more than on the initial stretch. This increase occurs because the spindle is stretched more than before. This condition is depicted in Figure 4–10 D. Clearly, the tension of the spindle muscle fibers, innervated by the gamma efferent fibers, sets the sensitivity and threshold for the whole spindle apparatus. From this organization, it is easy to understand some of the muscle physiology we experience daily.

The whole spindle apparatus keeps track of muscle tension. Not only does the muscle spindle monitor muscle tension, but also the

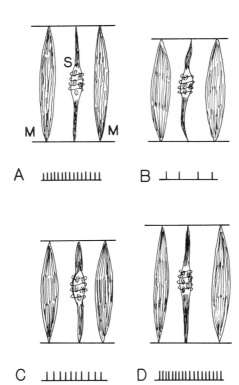

Figure 4.10. *Muscle spindle operation. The spindle (S) is parallel to the muscle groups (M). A. When the muscle group is stretched, the sensory firing rate (to the right of the letters) increases. B. When the muscle contracts, the spindle sensory firing decreases C. The spindle motor nerves increase firing and shorten the spindle to match the muscle length, restoring the sensory nerve firing to normal. D. Relaxing the muscle groups back to a normal length stretches the spindle, and sensory firing increases again, requiring the spindle motor nerves to decrease firing and restore the spindle muscle tone to normal.*

length of the whole muscle group. When the brain sets the activity of the gamma efferent fibers, it also sets such things as the body's posture. Muscles requiring more tension than others have more activity in the gamma efferent fibers. This corresponds to the set point in the general controller we looked at earlier.

Just as some "cross talk" occurs between hormonal systems, cross talk can also happen among the gamma efferent fibers. Rather general conditions such as anxiety can reset all the firing levels for the muscles, and both the general tone and sensitivity increase [28]. This cross talk may explain some of the hyperreactive reflexes often seen by physicians examining very nervous patients. Tension in other areas of the body can also add to an overall sensitivity. For example, if a person's hands are pulled apart while hooked by only one finger, tension reflexes in the knee and legs increase [29]. And out of this comes an interesting connection between a person who is farsighted and muscle tremor.

The whole nerve–muscle complex is a fast-reacting, servo system (a feedback controller following a changing set point) that is slightly underdamped [29]. In addition, it takes a finite reaction time for all the feedback events to take place. As a consequence, a muscle works in a slight "seeking mode" that generates a characteristic tremor of about ten cycles per second [29]. The amplitude of this tremor is determined by the spindle sensitivity, set in turn by the gamma efferent fibers. Gamma efferent fibers, on the other hand, are subject to influences from muscle tension in other portions of the body. For a farsighted individual, accommodating to near vision creates severe muscle tension in the eye muscles. As a result, gamma efferent fiber activity to the eye muscles is quite high. This elevated activity crosses into the other systems and the normal muscle amplitude increases and becomes quite visible [29].

The spindle system, however, is not the only system the skeletal muscle system relies on to gather information about the state of skeletal muscles. Located in the muscle tendon is a second sensor called the *Golgi tendon organ*. It is formed by the insertion of nerve terminals between twisted elements of the tendon's connective tissue [30]. This organ senses the overall tension in the tendons.

The basic organization of the reflex process is shown in Figure 4–11. Part of the input to the reflex system comes from the Golgi tendon organ. Between the tendon organ and the control system input is an interneuron that inverts the activity of the tendon organ from an excitatory state to an inhibitory state. This organization explains not only the pathways for the normal stretch reflex, but also a so-called inverse stretch reflex. Although the system shown in Figure 4–11 is quite simple, it produces the sophistication we experience with every muscle-mediated movement.

Because most of the control components are nerves, no requirements exist for special transducers to convert energy forms into the language of nerves. The system, then, needs only sensory nerves to acquire information. This contrasts with the hormonal controllers that require neuro-humoral transducers to maintain a relation between nerve and hormone activities.

The interconnection between the spindle sensory fibers and the nerves that control the remainder of the muscle forms the stretch reflex. Stretching the muscles causes the sensory nerves to increase

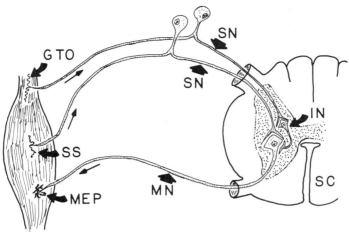

Figure 4.11. *Organization of the myotatic reflex. If muscle tension exceeds a threshold value, the muscle will suddenly release and relax to prevent damage to the muscle and tendons. The release is mediated through an interneuron (IN) acting through sensory information from the Golgi Tendon Organ (GTO) and a sensory neuron (SM). The spindle sensor synapes to a motor neuron (MN), extending to the motor end plate (MEP). SC is the spinal cord.*

activity and stimulate the motor nerves to the remainder of the muscle to contract and resist the stretch. This resistance is called *muscle tone.* If the muscle is rapidly stretched, greatly increasing the tension, the muscle will suddenly relax, and muscular tension goes nearly to zero. This "inverse tension reflex" is the result of the Golgi tendon organ, which inhibits the activity of the motor neurons through the inhibitory interneuron. With both systems functioning properly, muscles can maintain good muscle tone without damaging either muscles or tendons through an overdevelopment of tension. Still, if a muscle develops tension too fast for the control system, the resulting forces can rupture or tear tendons and muscles. Muscles that are chilled have depressed sensory nerve activity and are particularly subject to this sort of injury. "Warming up" brings the controls on muscle contraction and tone up to a proper operating temperature, protecting the whole muscular system from injury.

With this look into the neuromuscular system, some general conclusions about neurocontrollers fall into place. For example, neural sensors must receive a proper or adequate stimulus, which means that each sensor operates best when it receives energy in a form that it can best transduce. In addition, the sensory input need not come from a single sensor. For any given limb, information brought to the brain about the muscle condition can come from skin sensors, joint sensors, tension in opposing muscle groups, heat, cold, compression on the skin, and even the amount of blood flow reaching cells in the limb. And with all this sensory input, the data converge ultimately on the motor nerve that tells the muscle to contract.

In an earlier observation, it appeared that nervous systems had a low level of communications security and could cross talk to one another easily. This appears in some forms of muscle tremor or in the general massive reflexes that can occur in a paraplegic where one sensory input can stimulate the whole autonomic nervous system into action. The mass reaction can include urination, defecation, and wide swings in blood pressure [31]. Still, for all the cross talk that could occur and except for rare instances such as the paraplegic, most neurological control systems operate without seriously interfering with one another. Obviously, were this not true, the fine muscular control nearly all of us experience would be impossible.

The skeletal muscle system is a good example of a neuromuscular control system in the body. Another resides in the cardiovascular system, involving the autonomic nervous system and vascular smooth muscle. Because of its importance to body function, this system has received a great deal of attention. Let's take a look at what is known.

NEURAL CONTROL OF BLOOD PRESSURE

Just as in the skeletal muscle system, muscular control in the cardiovascular system depends heavily on the coordination among sensory nerves and motor nerves. The sensory input and motor output combine to form the reflex arc as in the skeletal muscle system. But the elements of the arc are more distant from one another.

Sensors that provide the appropriate transduction and input to the control system are located in the aorta and segments of the carotid arteries. Two forms of sensors are used, one to detect pressure, called a *baroreceptor*, and a second that responds to the blood chemistry called a *chemoreceptor*. Chemoreceptors are sensitive to the amount of oxygen and carbon dioxide present in the blood. Cardiovascular control, in an attempt to preserve blood flow to the tissues, detects the two main components of flow, the blood pressure required to move the blood, and the blood oxygen content delivered to the cells.

Sensory inputs of blood pressure and blood chemical qualities are used to regulate blood pressure through heart rate and blood vessel diameter (see Chapter 3 on the heart). These are the only two components of cardiovascular flow the body seems to monitor. As yet, researchers have found no natural "flow" sensors that detect and measure the volume flow of blood. The basic organization of the sensors and motor nerves is shown in Figure 4–12. Just as in the skeletal muscle

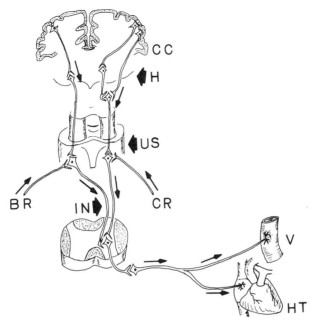

Figure 4.12. *Autonomic regulation pattern in the cardiovascular system. Sensory information enters the pattern through the baroreceptors (BR) and the chemoreceptors (CR). Excititory and inhibitory inputs come from the central nervous sytem, with BR inputs supporting inhibition (IN) and CR inputs supporting excitation. CC is the cerebral cortex; H is the hypothalamus; US is the upper spinal cord. The summed effects control the vessels (V) and the heart (HT).*

system, both excitatory and inhibitory elements are inputs to the control system. Let's step through the various controller components and see what each does in controlling blood pressure.

The motor nerves that send commands to contract to smooth muscle cells receive inputs from several sources including the brain's cortex. Other stimuli come from the respiratory center in the brain, with additional influences from pain, low oxygen, and higher-than-normal carbon dioxide in the blood. The chemoreceptors respond to a low oxygen or a high carbon dioxide blood content by increasing their activity. The neuro connection between these sensors and motor nerves is excitatory; that is, as the content of carbon dioxide increases and oxygen decreases, autonomic motor nerves increase in activity, narrowing the vessel diameters and increasing the heart rate, leading to a rise in blood pressure.

But decreasing vessel diameters and increasing the heart rate increases blood pressure, which stretches baroreceptors embedded in the aortic and carotid artery walls. This increase in stretch increases activity in the sensory nerves, but the connection with motor nerves is inhibitory. This sort of antagonistic control from two different inputs to the single controller is a familiar pattern found earlier in hormonal controllers.

The various inputs to the autonomic motor nerves shed some light on how certain events occur when the body is stressed. For example, a low blood oxygen level will increase the heart rate and blood pressure in an attempt to increase the amount of blood reaching the tissues. But just when the pattern seems obvious, the simple associations begin to fail. Pain normally increases both blood pressure and heart rate. Chronic pain, on the other hand, can cause massive vasodilation and blood pooling away from the heart and fainting. Clearly, we are not seeing all the control mechanisms in this model.

Nevertheless, the model does answer many of the control functions measurable in the cardiovascular system. For example, the combined stimulatory and inhibitory inputs to the heart present an opportunity for a control system "hunting mechanism" to appear, where the time delays through the controller cause a series of small oscillations in the controller output. And indeed, such oscillations appear in long-term blood pressure measurements [32]. The oscillation rate is about one cycle every 20 to 40 seconds, and the oscillations become quite evident during periods of hypotension, for example, when the cardiovascular system experiences shock.

The stimulatory input to the heart and vessels comes from a portion of the autonomic nervous system designated "sympathetic." These nerves have the common property of using norepinephrine as the neurotransmitter. This neurotransmitter directly stimulates the heart to increase its frequency of contraction and also to increase its strength of contraction. Additionally, smooth muscle in the vessel walls contracts in response to norepinephrine.

The inhibitory input to the heart, in contrast, is the parasympathetic nervous system, a direct antagonist to the sympathetic system, using the neurotransmitter acetylcholine. As a consequence, the heart sees two mutually antagonistic inputs from the sympathetic and parasympathetic systems. It is the balance of these two control mechanisms that ultimately sets the final functioning level of the heart.

The heart is not the only organ with an antagonistic innervation from these two systems. The list is a real "who's who" in organs and includes: the iris of the eye, the tear glands, the salivary glands, the bronchial tree and vessels in the lung tissue, liver, bile ducts and gallbladder, spleen, gastrointestinal tract, kidney, urinary bladder, and the genitals. A glance over the list shows an interesting common denominator: all these organs are associated with functions we normally consider outside our conscious control. (We do have some control over urination and defecation because one of the two sphincters on each of these outflow tracts is consciously controlled.) Still, evidence indicates that centers in the cerebral cortex, the rational portion of our brain, extend nerve fibers into the two autonomic divisions that control the heart [33]. Similar control-links from the higher portions of the brain can be found for other organs as well. Clearly, then, the question remains: Just how much influence does the cortex of the brain have on the autonomic system? The answer may be: much more than suspected. We are just beginning to learn how much.

The turning point in this notion of higher controls on the autonomic nervous system came in the late 1960's when a researcher demonstrated that rats could be taught to control blood pressure, rate of urine formation, and a number of other processes that fall under the heading of automatic [25]. Experiments in biofeedback suggest that humans could do the same sort of thing. The more successful practitioners of biofeedback can make one hand warm and the other cold, seemingly on command. Blood pressure and heart rate can be lowered on command. Some of the Yogis of India seem able even to lower their cellular metabolic rate. The medical community is just learning what all this can mean in the treatment of disease.

CONCLUSIONS

A careful glance at the living things that surround us shows that control is an essential part of living. Without control, the dedicated activities of the cells and organs could not be directed toward the benefit of the living organism. Although we have emphasized the human in our discussion, these controls are mimicked in living things from man to worm. Control keeps things working in concert to keep living things living.

For the more complex animals including man, control can be quite sophisticated, utilizing two means of communication, nerves and hormones. They often work in coordinated units, but may also be specialized, dealing with just hormones or just nerves.

Controls through hormones rely on molecular shape as a specific message that can be carried by the circulation to remote parts of the body. The molecule, because of its stereochemistry, can be quite specific about the message it carries. With hormones, the message is the medium. And that message can reach all the cells in the body, with all of its power and uniqueness.

Nerves, originating in the central nervous system, reach out sensory and motor nerve fibers throughout the body. Messages carried by nerves can be delivered quickly, but they carry some chance for cross talk into other neuro patterns. Still, nerves are fast, bringing messages coded in the frequency of action potentials and finally handed to the innervated cells by the neurotransmitters. Nerves also operate in hormonal control systems where many of the primary instructions that pace hormonal secretions come from the brain. The rhythms may span hours, days, months, or even years.

Nested at the base of the brain is the master gland, the pituitary. It receives information and instructions in the form of chemicals from the brain and transduces these messages into a form the rest of the body can understand. The influence of the pituitary on the body seems all-encompassing. When the gland malfunctions or is disrupted by an invading tumor, the outward signs can be quite dramatic. The day-to-day signals emanating from this gland pass through the vascular system to regulate gland secretions that in turn regulate great sections of body function. Although a cliché, the term "master gland" carries the right meaning for the pituitary.

The search for pure neuro controllers outside the brain was not successful. We had to examine instead nerve–muscle controllers. They had a form quite unlike the organization of the hormonal controllers, and depended upon the nerve action potential to carry the message. Indeed, it may appear that the action potential is everywhere. The event is so central to all that happens in the body, it simply cannot be treated once and forgotten. It is testimony to the action potential's cardinal position in biological systems.

In the end, we have viewed only a few of the multitude of controllers, known and unknown. The body is like a huge jigsaw puzzle of interlocking controllers, with edges that fit so tightly that tracing the fit is nearly impossible. Despite our limited view, we can stand back for a moment and vaguely sense the "busyness" of all the controls that govern the activities of all our cells and ourselves.

5

GETTING THINGS INTO AND OUT OF THE BODY

The body is not isolated from the outside world; it is a part of it, a part of the main, to paraphrase John Donne. A very active exchange of molecules passes between the inside and outside of the body. In previous chapters, we observed that cells required the right sort of environment to live, and many of the body's functions are directed at maintaining this environment at a correct temperature, with a correct amount of water, and with a correct mix of molecules for energy and to make tissues. Somehow, water and other essential molecules need to move into the body, while discarded molecules need to move out. Through these interfaces and their control mechanisms, internal cells are bathed in fluids and molecules central to life.

Yet, it is not so obvious why hypodermic needles are still part of medicine when so many pills are available and are so much easier to take. To understand this dilemma, let's ponder the case of penicillin for a moment.

Perhaps the most obvious route to the inside of the body is through the skin. Rubbing penicillin on the skin, however, will not get it into the body because the skin is a very effective barrier. On the other hand, we might inhale penicillin into the lungs as a fine powder, much like smoking a cigarette; but after the sneezing and coughing, not much would be left in the airways. In addition, any remaining molecules would likely be scooped up by busy white blood cells that stand guard at the lung–air interface. We might eat penicillin, but the acid in the stomach would quickly inactivate the molecules, leaving nothing effective to enter the bloodstream. (Some of the later versions of penicillin are acid-resistant, however, and can withstand stomach acidity.) We might try an enema, but in our society that route of administration is viewed with little enthusiasm. So, if we need penicillin to ward off a life-threatening infection, it often comes by hypodermic needle.

In the case of penicillin, the list of attempts to get this substance into the body points out the obvious interfaces that are sites of molecular exchange with the outside world. Considering only the functional position of the interfaces, the body can be simplified to Figure 5–1. Although the diagram looks very little like a human, it nevertheless shows the body's various portals to the outside world.

In Figure 5–1, the complicated and convoluted gastrointestinal tract is simplified to a long tube. At the same time, the cardiovascular system becomes a flexible tube loosely draped about the interior of the cylinder. Lungs and kidneys have become infoldings of the surface. Notice that no direct opening to the outside world exists through any of the interfaces. A surface of cells always protects the interface, forming a separation between bloodstream and outside world. In addition, the interface is folded into the body away from the external parts of the body. This design is an effective way of protecting the interface

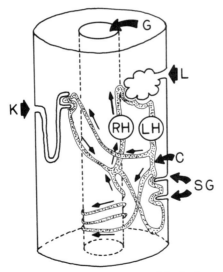

Figure 5.1. *An organization of the body showing the various interfaces with the outside world. The body forms a dynamic system openly exchanging with the outside world. G is the gastrointestinal tract; L is the lungs; K is the kidney; SC are skin glands; C is the circulation; RH is the right heart; LH is the left heart.*

from bacteria or larger organisms that could easily penetrate through a layer of single cells.

The layers of cells between the blood and the outside world set up a distinction between inside and outside the body. For instance, being inside the gastrointestinal (GI) tract is not being "inside" the body (see Figure 5–1). Getting inside the body from the GI tract requires penetrating the walls of this structure. Thus, something swallowed is not inside the body and will not be until that something finally passes through the walls of the tract.

This organization is not far from the basic worm tunneling its way through the soil of a green lawn somewhere. Just like the worm, as the food "passes by" in the intestine, we pull out the molecules needed and leave the rest.

Interfaces can take on different forms. For example, an interface may be like the intestine, where local processes remove specific materials and leave the rest. In contrast, an interface can be like the kidney, where an elaborate process produces a complex fluid, which is finally excreted from the body. This is a one-way process, always out of the body. It is not common to think of putting something into the body through the kidneys. The lungs, on the other hand, exchange in both directions and offer a way into the body. The skin, however, is an unusual interface that is both watertight and yet heavily involved with water loss from the body. We will look at all of these interfaces in more detail, examining those features that make each interface unique and those that are common to all interfaces.

THE SKIN

Humans place a lot of social importance on the skin. It provides most of what is seen of another person, so, when it has a pleasant texture, the person is usually considered attractive. Most of what is seen is dead, however. Covering the outside portion of the skin is the *stratum corneum* or so-called horny layer of the skin. This layer is composed of dead cells, each filled with a hard substance called *keratin*. These cells begin life as cuboidal cells deep in the germinal layer of the skin, then slowly migrate to the skin surface, changing shape and internal composition in the process and finally dying. They are shed in huge numbers every day, and each cell that is cast off must be replaced, so the rate of cell replacement in the skin is very high. We lose these tiny flattened plates with each movement, every small friction against the skin from clothing, every loving caress from another. As we walk through life, dead cells leaf-off behind, leaving a little of ourselves that eventually becomes part of the environment.

Projecting through the dead cell layers are two types of openings that secrete substances onto the surface of the skin. The first is a duct from a set of glands called the *eccrine* glands. They secrete a weak saline solutioin with a small amount of nitrogenous material in it [1]. The second is a hair follicle, through which projects a hair. The follicle secretes a material called *sebum*, a solution containing high amounts of lipids and fatty substances [2]. A schematic drawing of the skin and its glands is shown in Figure 5–2.

The sweat glands and hairs are unevenly distributed over the body, varying both in size and number in various skin areas. At the same time, the texture and thickness of the skin vary over the body, ranging from the thick layers of the soles of the feet to the delicate, thin layers of the eyelid. The armpit has coarse hair and a group of sweat glands called *appocrine* glands [3]. The appocrine glands often occupy the same skin area as hair, and the dimensions of the glands can become quite large, ranging from 0.3 mm to 0.5 mm in diameter. This variability in texture and composition of the skin is not random, but reflects working requirements for each skin area or a response to use.

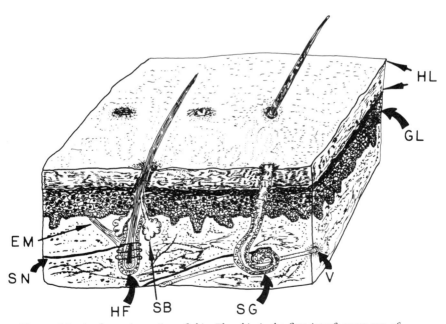

Figure 5.2. *A schematic section of skin. The skin is the first interface we see of another person. On the outer or horny layer (HL) are dead cells, produced by the germinal layer of cells (GL). Penetrating through the skin are hairs with sebaceous glands (SB) and sweat glands (SG). The hairs have erecting muscles (EM) and a follicle (HF) that makes the hair. Around the follicles are sensory nerves (SN). Vessels (V) travel close to the surface to transfer body heat to the outside.*

Materials secreted by the skin can vary; for instance, the ear has skin glands that produce a fatty-based wax that covers the ear canal [4]. Still other examples are the bottoms of the feet and the palms of the hands, which are hairless and covered with a different type of sweat gland. Most sweat glands operate as part of the body cooling process, which will be examined later. Finally, moisture, warmth, and fatty substances exuded onto the skin by the many skin glands create an attractive environment for another set of organisms, bacteria.

Stuffed into corners and niches of the layers of dead skin cells are bacteria. Along with the bacteria, a few even more primitive organisms called *fungi* reside, but normally not very many. The bulk of the population is bacterial, and they are from two major bacterial populations. First is the group called *cocci* with spherical cells visible under a microscope [5], and second is the *diptheroid* group with irregular rod-shaped cells [6]. Most of the cocci are harmless, except for the staphylococcus group. *Staphylococcus aureus* causes pimples, boils, and some serious infections that suddenly appeared in hospitals during the early 1960's. *Streptococcus B*, on the other hand, causes serious infections in newborns, infections that can lead to death or defects such as blindness or deafness [7].

Under normal circumstances, bacteria on the skin cause no serious problems and are seldom a threat to life. When the body is opened during surgery, however, bacteria can be a source of lethal sepsis (infections), and the aseptic procedures that are part of modern surgery attempt to keep bacteria from entering the body through a wound.

In general, the distribution of skin bacteria takes on a profile that follows the best environment for the bacteria. For example, portions of the skin with a large number of sebaceous glands (the skin glands that secrete sebum) host *anaerobic* bacteria (those that do not use oxygen) in greater numbers than *aerobic* bacteria (those that do use oxygen) in a ratio of about 10 : 1 [8]. The distribution depends upon a bacterium's ability to use fat as an energy substrate. It is quite evident, however, that for each individual the distribution is as singular as the person, and the distribution is rather stable over time. New arrivals have a tough time getting a foothold on the skin when all the available living niches happen to be taken by bacteria already living on the skin.

The steady-state turnover of bacteria on our skin can change, however, if an agent such as an antibiotic should kill off the residing bacteria. On the heels of the old bacteria dying, new ones can move into now-available niches. These new organisms, however, may not be as innocuous as the previous occupants. Often the new inhabitants are fungi that become parasitic on the body, causing tissue damage [9]. These are the so-called super or secondary infections that follow the use of broad-spectrum antibiotics.

Bacteria can occupy the outer surface of the skin because moisture and warmth are there, and for the skin, these two qualities are quite intimately connected. Because the environment outside is usually cooler than the body's interior and the skin lets heat flow in either direction through the skin, body heat flows to the outside, and the body constantly makes more heat to replace that which is lost. If the temperature inside the body should rise above normal, the skin and cardiovascular system combine to increase heat flow to the outside. The skin then becomes one of the heat-regulating elements of the body.

Just below the skin are large networks of capillaries controlled by local vascular beds. Bringing blood to the surface of the body in large amounts increases the amount of heat transferred to the outside. In this case, the circulation transports heat from deep within the body to the body's surface. On the other hand, the vascular system can conserve body heat by decreasing blood flow to these capillaries, reducing heat loss. The flow of blood close to the skin adds color, so when the capillaries are contracted, the skin looks blanched, causing the chilled face of winter. In contrast, when heat is being removed from the body, an increased blood flow to the skin causes the ruddy complexion of an exerciser. If the skin has a natural coloring, some of the changes will be hard to see, but they still happen, only hidden by the pigmentation.

Heat flow through the skin follows the rate of heat removal at the skin surface. As a result, increasing heat loss from the skin surface increases total body heat removal. Now, the sweat glands can play an active role. As the sweat glands excrete fluid onto the surface of the skin, the fluid evaporates, taking 540 calories for each gram of fluid evaporating. In the course of a not-so-busy day, we can excrete about 50 milliliters per hour as an insensible water loss [10]. Sweating goes on even though we do not feel it, carrying off some 27,000 calories of heat every hour.

If a person happens to be nervous, excited, or fearful, sweat production follows a different plan in terms of quantity and location. Sweat appears on feet and hands (the famous sweaty palms), under the arms and in the crotch (no famous name for that one). All of these are confined spaces except for the hands, and bacteria grow rapidly in response to all the available food and water. The results are all the body odors we try to wash, spray, and perfume away. These bacteria, lodged on and in the skin, keep a lot of people busy coming up with chemicals that will either hide the odors or slow sweat gland secretion. A whole industry exists to deal with these small organisms invisible to the naked eye.

Despite the water that may cover the skin when a body is overheated or nervous, the *squamous* cells that form the horny layer continue to make the skin nearly waterproof. How essential this feature is to normal life becomes evident when burns or trauma remove the skin and an attending physician must work hard to stem a steady loss of body fluids. Burns over large portions of the body remove protective layers of skin, opening the door to serious infections and lethal losses of body fluids. But the skin does more than keep inside water inside; it also keeps outside water outside. A body is not faced with a water balance problem if it should be caught in a rain, take a shower, or soak for a while in a hot bath. And along with this integrity, we can walk, reach, and stretch because of this thin, flexible, and renewable covering called skin.

Although waterproof, the outer skin is not chemical-proof. Some chemicals can penetrate through the skin easily, especially if the material happens to be lipid-soluble (it can go into solution in fats or oils). Lipid solubility is an essential natural property of any material that can penetrate lipid-rich cell membranes. One such substance is DMSO, an agent with unusual penetrability. For example, if DMSO is applied to the skin of the fingers, it can be tasted in the mouth a few seconds later as the bloodstream carries it to the taste buds on the tongue. Other drugs can be carried by agents like DMSO and quickly penetrate the skin along with the carrier, potentially replacing the hypodermic in speed and efficiency. These penetrating agents are under study not only for an ability to carry another molecule through the skin, but also for any secondary therapeutic value the carrier may have in such areas as inflammation, allergy, or even cancer.

Portions of the skin are specialized for transporting molecules through the interface, but this is not a property of the skin only; it is also an attribute of the "inside" skin, the covering on the intestinal tract. Because of our consideration of what is or is not inside the body, it is not surprising to learn that the GI tract lining and the outside skin have similar functions. The skin on the GI tract, however, does not have the dead cell layer of the outer skin, although it has nearly the same rate of cell loss [11]. The GI tract will appear in more detail later in the discussion.

Along with heat and water, a third form of energy crosses the skin: sensory information. Although the skin has the smallest area of all the body's interfaces, ranging from 0.86 m^2 to 2.8 m^2, it positions a dense population of sensors at the extreme exterior of the body. Heat, pressure, cold, chemical, and stretch sensors are distributed over the body just under and in the skin. In summary, along with the skin's function as a barrier to fluids moving in either direction and a barrier to bacterial invasions, it also provides sensory information and a regulation of heat flow.

In contrast with the skin, the lungs are a much busier interface, yet sharing some of the properties of skin.

THE LUNGS

We are probably less aware of the lungs as an interface than of the skin because the lungs are positioned away from the body's exterior, inside the chest, where they cannot be seen. Nevertheless, they form an active interface for the exchange of oxygen and carbon dioxide. Because the interface is located some distance from the opening that conveys air into the chest, breathing must bring oxygen into the chest to the interface. As oxygen reaches the interface and moves through it, carbon dioxide moves across the interface in the opposite direction, and must then be expelled through the airway space.

The physical breathing process we go through about 14 times each minute moves oxygen in and carbon dioxide out by physically moving air into and out of the airway space. Airways into the site of gas exchange are a branching system of channels much like the branches of a tree. The trunk of the tree corresponds to the *trachea*, with the limbs and branches of the tree corresponding to the *bronchi* and *bronchioles*, the smaller branching channels of the lung. At the end of the branching processes are the *alveoli*, which are the sites of gas exchange.

The lung is formed like a blind pouch, requiring the physical motion of air into and out of the pouch in a breathing cycle to complete the exchange of gasses through a rather complex system of airways and sacs.

Air enters and leaves the lung by physically changing the lung's volume. Changing lung volume comes about when the chest cavity changes volume. The volume change, in turn, comes from the contraction of a large muscular sheet that is stretched over the bottom of the chest cavity. This muscle is called the *diaphragm*. In addition, the chest can increase volume through the contraction of the chest muscles. A contracting diaphragm increases the thoracic volume when its normal dome shape becomes flattened during contraction. Chest muscles increase the thoracic volume by pulling the ribs outward.

The ability of the lung to follow a change in chest volume depends upon a thin cavity called the *pleural cavity* that surrounds the lungs. The cavity is not a real cavity at all, but a double-sided sac that surrounds the lung. One side is attached to the lung, the other side is attached to the chest wall and diaphragm. Inside the pleura, the pressure is less than the outside atmospheric pressure. So, as the chest expands with inhalation, the lung tissue travels with it, coupled by the subatmospheric pressure in the pleural space. The whole arrangement is shown in Figure 5–3.

When taking a breath, the airways convey air but do not take part in the gas exchange, and because of this, air drawn in by a contracting diaphragm does not end up with the same gas mixture at the alveoli

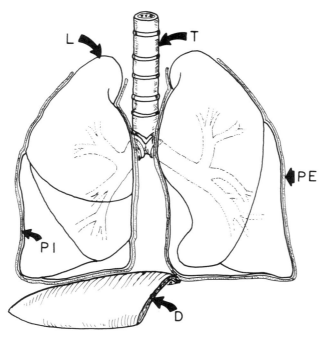

Figure 5.3. *Organization of the pleural sacs, diaphragm, and lung. The pleural sacs surround the lungs (L) in a blind fold, forming an internal pleura surface (PI) and an external surface (PE) that is really a single complete surface. A small amount of fluid between the two surfaces and a subatmospheric pressure keeps the sac functionally closed. The diaphragm (D) is a dome-shaped muscle that contracts to expand the lungs through the pleura. T is the trachea.*

as the gas mixture outside the mouth. To start with, inhalation brings air into the lung, heats the air, and saturates it with water, reducing the partial pressure for all the mixture components. The air then mixes with the remaining gasses from the last expiration. All of this mixing, first with water vapor and then with residual gasses, lowers the concentration of oxygen at the alveoli.

The volume of air remaining in the nonabsorbing airways is called the dead space. This dead space can increase by attaching longer tubes to breathe through. Because the lung is a blind sac, increasing the dead space reduces the oxygen reaching the alveoli. In extreme cases, breathing through a tube with a volume larger than the changes in lung volume prevents inhaled oxygen from ever reaching the lung interface. If the volume is made slightly larger than normal by adding a short breathing tube, the dead space will contribute to the retention of more carbon dioxide in the lung and in the blood. The total pulmonary space that results from putting someone on a respirator creates secondary problems for anesthetists and therapists. Respirators must be designed to limit the dead space added to the whole respiratory system.

The actual site of gas exchange is the alveoli, which are grouped into several thousand clusters that make up about 300 million small sacs in the adult. Each cluster is surrounded by a dense set of capillaries. Also weaving through the vascular network is an extensive network of lymphatic channels. The lymphatic system has two identified functions in the body: to provide part of the immune response of the body; and to remove fluids from extravascular spaces. In the lung, both functions of the lymphatic system are quite active.

Like the lymphatic system that can respond to the presence of fluids or bacteria, the vascular system surrounding the alveoli will respond to the mixture of gasses. When a nearly normal level of oxygen is present in a group of alveoli, the pulmonary vessels surrounding the alveoli expand, increasing blood flow through this oxygen-rich portion of the lung [12]. In contrast, these vessels will close with abnormal levels of carbon dioxide [12]. Such a response prevents the lung from setting up poor mixes of oxygen-rich and oxygen-poor blood. A potential for poor mixing occurs when pneumonia reduces oxygen flow through a portion of the airways. Blood flow through the poorly ventilated portion of the lung will decrease as the amount of carbon dioxide increases and oxygen decreases in that region.

In the alveoli, oxygenation occurs by the diffusion of oxygen across a thin layer of cells that separates gasses from the blood. This layer is so thin that certain molecules can quickly diffuse through it into the bloodstream. Bacteria are also able to penetrate this barrier, so the lung has an elaborate immune system to keep at bay any bacteria that enter with the air. Because this cell layer is so thin, the lung is a potentially "easy" route to administer drugs. The initial requirement is that the drugs be water-soluble to penetrate through a thin layer of fluid covering the lung lining.

But the biggest problem of getting drugs through the lungs is the automatic coughing and sneezing response of the respiratory system to irritants. Irritating the airway walls with particulates or a drug will set up the coughing reflex. In this process, the chest muscles contract to rapidly build up pressure in the airways. Suddenly releasing the pressure by opening the glottis forces the air through the airways at velocities exceeding 500 miles per hour [13]. These velocities usually expel most materials that gain entrance to the airways.

Organisms and substances not blown out by a cough are permitted to settle on a "river of mucus" secreted by special cells along the airway. This mucus, with its captured materials, slowly moves toward the trachea and eventually out of the lung. Moving the mucus are cell appendages called *cilia*. These cilia are extensions of the cell that move using contractile proteins like those found in muscle. By contracting these proteins, the cilia move about. When millions of these cilia-containing cells are brought together and coordinated to work in groups, the result is a wall of moving, hairlike appendages that slowly sweep the mucus out of the lung. Caught in the mucus are bacteria, viruses, cells dislodged by coughing, white cells attacking bacteria, and particulates. The mucus takes on the color of its contaminants; in the case of a smoker or coal miner, for instance, it darkens. The mucus and cilia combine to form an effective means of capturing unwanted substances and moving them out of the lung.

If the bacteria and particulates should get by the cilia and reach the deeper parts of the lung, then highly mobile white blood cells ingest the smaller foreign bodies. When the particles are dark, they combine to darken the lung's appearance. As a result, smokers, miners, and inhabitants of large cities have darker lungs from materials they retain within the lungs.

A dark lung is not the only problem that can plague a cigarette smoker, however. Carbon monoxide, a major component of cigarette smoke, has more than one biological effect. For example, beyond its ability to replace oxygen in the hemoglobin molecule, carbon monoxide also has an anesthetic effect on mobile white blood cells [14]. As a result, carbon monoxide-exposed white cells are less able to respond to the signals of an infection. In addition, this anesthetic effect extends to cilia-containing cells, and the cilia decrease their sweeping motion, leaving additional mucus deep in the airways. When the build-up of mucus is large enough, it causes a cough reflex, and the smoker has a "productive" cough, bringing up mucus in clumps with each coughing episode. Thus, carbon monoxide has not only opened the door to infection, but also decreased the lung's ability to clear out foreign matter. It is no surprise that cigarette smoking brings an increased incidence of lung infections, both viral and bacterial.

Bacteria are not the only things that can gain entry to the lung with smoking. Along with the smoke are gasses and tars that contain carcinogens and precursors of carcinogens. These agents penetrate the lung's cells and there they function as cancer-causing agents or promoters, making the cell vulnerable to other cancer-causing agents [15]. Cells that are exposed to these agents physically change and appear under the microscope as precancerous.

But the lung is very resilient and is often able to recover from the constant insults of smoking. Because of normal cell death and replacement, old, sick cells are removed and replaced by new ones when a smoker quits. Yet, even with the stress of cigarette smoking or working in a dust-laden environment, the lungs perform their task, bringing in oxygen and removing carbon dioxide with each breath.

The lungs have a large vascular system that can hold a sizable blood volume reserve. And the serial position of the circulation makes it an effective site for introducing special biochemical functions. For example, the lungs supply an enzyme that converts angiotensin I to angiotensin II, delivering one of the most potent cardiovascular hormones to the body. Like the skin, the lung has many roles in the body.

Are the lungs organized in the same way in all animals? Mammalian lungs are organized like ours. On the other hand, insects have a simple lung that provides an exchange interface for oxygen and carbon dioxide. Birds, however, are different in a curious way. Instead of the blind sac for air flow, a bird's lung has many small tunnels surrounded by capillaries. The air passes through the lung in one direction only, while blood flows in the opposite direction through the capillaries. As a result, birds are able to extract more oxygen from the air on each breath than we. This highly efficient lung organization helps birds fly as high as 20,000 feet above sea level while expending large amounts of energy in flight.

Whether bird, insect, or human, so long as the body uses oxygen as part of its energy cycle, oxygen must enter the body and reach the cells. And in each case, we find the lungs provide the interface while the circulation provides the delivery system.

Oxygen is only part of the show, however, for cells need other molecules to produce energy, to replace old cells, and to rebuild body components. These molecules enter the body through the next larger area, the gastrointestinal tract.

THE GASTROINTESTINAL TRACT

The gastrointestinal or GI tract is the center cylinder in Figure 5–1. Before the physiology of the GI tract was understood, its role in human affairs was subject to all sorts of interpretation. The ancient Egyptians viewed the anus as the seat of putrefaction and disease [16]. They went to great lengths to clean, bathe, and soothe the opening and thereby prevent the onset of disease. This may have been the first attempt to prevent disease through a prescribed set of activities, what we now call preventive medicine.

Watching the bowels and their movements became a great attraction during the seventeenth and eighteenth centuries in Europe. Louis XIV was reported to have experienced no fewer than 2000 enemas during his reign [17]. In the new world, the Mayans used the enema for the far different purpose of inducing intoxication. Because of the physiology involved, the Mayan practices will be covered in more detail later. The GI tract, despite all this attention through the centuries, is a mystery still being unfolded. With new, complicated hormonal feedback systems being discovered, the GI tract is taking on a high level of sophistication. A diagram of the GI tract is shown in Figure 5–4.

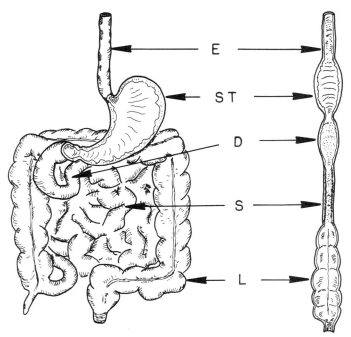

Figure 5.4. *The basic elements of the gastrointestinal tract. The tract begins at the esophagus (E), which empties into the stomach (ST). A specialized segment of the small intestine (S) is the duodenum (D), which then empties into the large intestine (L).*

In reality, the GI tract has several functions that are well understood. They include: the preparation of food for digestion; storage for the digestion process; supplying agents and conditions for digestion to occur; providing an interface through which molecules can reach the bloodstream; and providing a dehydration that finally reduces the residue of the eaten materials into feces. Let's examine each of these functions along the sequence that starts with chewing and swallowing and ends with semisolid feces.

Preparing food for digestion is mostly a matter of chewing. This simple process breaks and grinds the food into small pieces that have a large surface-area-to-volume ratio. Mixed in with the chewed food is saliva that softens and lubricates the food for swallowing. Within the saliva are a few digestive enzymes, so digestion technically begins in the mouth.

From the mouth, the chewed food moves to the stomach through the *esophagus.* The stomach provides temporary storage for the food during the digestive process and is a source of digestive chemicals. The stomach has two valves or sphincters located at opposite ends of the organ. The *upper* or *cardiac valve* is located at the junction between the esophagus and the stomach; the *pyloric valve* sits at the outflow opening, at the junction between the stomach and the *duodenum*.

The cardiac valve is often poorly defined, with only a small amount of sphincter muscle. The pyloric valve, in contrast, is very well developed, with large annular rings of muscle to close or open the orifice on command.

When food is in the stomach, cells in the stomach wall secrete enzymes and large amounts of *hydrochloric* (HCl) *acid*. The acid, it turns out, directly contributes very little to the digestion process. Instead, the acid lowers the pH and the free hydrogen ions activate the enzymes that carry out the actual digestion. The enzymes, working as biological catalysts, speed up the dissociation of food molecules. For example, large protein molecules are digested into amino acids used later to make cell proteins. The small pieces of food produced by chewing expose a large amount of the food to digestion at the same time. The chewed food is dismantled as the acid-enzyme system breaks up large molecules into more fundamental components. These fundamental molecules are absorbed into the body and become an internal source of energy or parts of new, more complex molecules for a wide range of biological activities.

Although the stomach can secrete acid and enzymes to digest food, it does not digest itself along the way. This is a unique property of the stomach wall, for the walls of the intestine can be digested. If the stomach wall protection should vanish, the stomach would start to digest itself. The result is a gastric ulcer.

When the acidity of the digested food reaches the proper level, the pyloric valve opens, and the stomach's contents, called *chyme*, moves into the duodenum. Beginning in the duodenum and extending over the remainder of the small intestine is a new wall texture and organization. Finger-like projections called *villi* reach into the lumen of the intestines, and on these villi are even more projections called *microvilli* (see Figure 5–5). The microvilli are about 0.5 mm to 1.5 mm in diameter, each providing about 10 to 40 mm^2 of area. These projections greatly multiply the effective surface area of the GI tract without greatly increasing its length.

In total, the small intestine is about seven meters long after death, but about three meters long (a little more than nine feet) in life. This difference in length comes from the relaxation of the smooth musculature that squeezes and propels the food through the intestine. A layer of contracted longitudinal muscle deep within the intestinal wall shortens its overall length in life.

Along its three-meter length, the small intestine is divided into three distinct segments, each with special absorption properties. The first section is the duodenum, named for its length, about 12 fingerwidths long (from *duodeni*, meaning twelve each). That turns out to

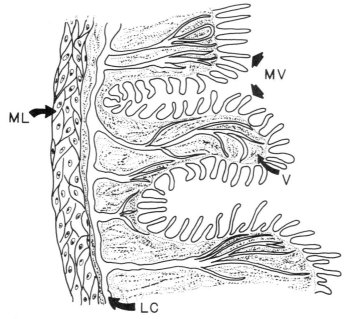

Figure 5.5. *The villi and microvilli of the gastrointestinal tract. To increase its effective area of absorption, the gastrointestinal tract uses villi (V) and microvilli (MV). A muscle layer (ML) coats the tract, and extensive lymphatic channels (LC) permeate the internal villi.*

be about 25 cm. Vascularization into the duodenum is very perfusive, suggesting an extensive absorptive role for this small length of intestine. The remaining sections, the *jejunum* and *ileum*, have a combined length of about 250 cm (100 inches). The transition between the duodenum and jejunum is much more distinct than between the jejunum and ileum. Some bacteria reside in the ileum, but not as many as in the colon, which is later in the sequence.

Despite the apparent shortness of the small intestine, it provides nearly $300 \, m^2$ of absorptive area, about the size of four badminton courts, or four times the area of the lungs and 200 times the area of the skin. It is the largest interface in the body.

At the end of the small intestine, the colon begins with a structure called the *cecum* and a small appendage forming a blind pouch. This small finger of tissue is the appendix. In the colon, a new set of processes begin that represent the last stage of extracting water and molecules from the material passing through the GI tract. The colon or large intestine also has extensive vascularization, but lacks the microvilli characteristic of the small intestine. The colon has ascending, transverse, and descending segments, ending finally in the sigmoid colon that communicates with the outside through the anus. The whole colon averages out just a little longer than one meter. In this meter, the final extraction of water and a few additional molecules like amino acids and some vitamins occurs.

Water extraction in the colon is quite effective. For example, the colon receives about 300 to 500 ml of chyme from the small intestine daily, which the colon dehydrates down to about 150 grams of semi-solid feces.

In the colon is a rich variety of bacteria. They generate a small amount of gas that when added to swallowed air forms flatulence. Along with gas production, however, the bacteria produce vitamins, which are absorbed and used by the body.

For the GI tract to absorb materials like vitamins and amino acids, the products of digestion must contact the intestinal wall. Peristaltic waves of contraction propel food down the GI tract. Although the GI tract can carry on without a supply of nerves, the sympathetic and parasympathetic nervous systems have profuse nerve endings into the smooth-muscle walls of the tract to regulate the movement of chyme. As a result, drugs that look like neurotransmitters can speed up or slow down chyme movement through the lumen, which can be the primary purpose of a drug or just one of its side effects.

Transit time through the GI tract gives some measure of how fast chyme moves along, and how long it takes for a meal to finally clear the body. From ingestion, it takes about four hours for the leading segments of the meal to traverse the small intestine, with the remainder following thereafter [18]. In about six to nine hours, the meal is almost completely in the colon. Then it takes six to nine hours to traverse the flexures of the colon, placing the meal into the sigmoid colon in about 12 hours. After 72 hours, about 25 percent of the meal remains in the colon. Other experiments show that about 70 percent of the meal can be recovered in about 72 hours [18]. Total recovery takes more than a week. The movement pattern is simple; fast in the small intestine, and slow in the large intestine, setting up first rapid absorption of molecules and later a slower absorption of water.

Most of us, however, are not too worried about how long it takes for food residue to make the trip through the GI tract, except for two conditions. First is an inability to get things moving, called constipation. Second is an ability to get things moving too well, called diarrhea. In moderation, neither of these conditions poses any real threat to the body. In extreme, diarrhea can be life threatening.

Constipation can be a medical problem if it lasts for more than a month. In some areas of the world, a condition called *megacolon* occurs, and the GI tract fails to move residue through the colon, forming a very large fece called a *fecalith*. With time, the fecalith grows large and hard, causing no serious problems or toxicity. When the fecalith becomes too large, it is surgically removed. That is constipation in the extreme. In the milder cases of constipation, we may feel some discomfort, which comes not from the supposed toxicity of the feces, but from a rather complex neural reflex that begins with neuro sensors in the rectum. The discomfort of constipation can be completely reproduced by replacing inert materials in the rectum [19]. So the unpleasant feeling of irregularity is a good indication that our nervous system is operating quite well, and if the constipation should continue for a while, we are in no real danger of "self-poisoning."

In contrast to constipation, diarrhea can be a problem because body fluids and essential minerals are lost in the process. Even moderate diarrhea that lasts for a long time can cause measurable deficiencies of essential body minerals such as potassium and sodium. Diseases like cholera that cause a massive flow of body water into the colon in a very short time are lethal without therapy to support the body's fluid balance [20]. In the absence of replacement fluids and antibiotics, cholera would remain an untreatable disease. In more remote parts of the world where such therapy and immunization are not available, cholera carries the same frightening consequences and generates the same fear as when the disease appeared in wagon trains moving to the American West in the 1800's. In the more profound forms of cholera, as much as 25 percent of the body's water can be lost in a matter of hours. That much water loss is almost uniformly fatal. But most of the diarrhea suffered in the U.S. is not so serious.

The colon conserves body water by extracting water from the stored food residue against a water concentration gradient. This requires active pumping of water into the body and opens a means of rapid access to the body's circulation for agents such as anesthetics, sedatives, tranquilizers, and steroids. Most of these agents can cause some gastric upset, such as nausea, vomiting, or diarrhea. Applying these agents through the rectum, however, avoids the usual gastric upset, and these substances can enter the body in proper amounts.

We are not the first to observe how quickly things can get into the body by this route. The Mayans used a group of over 25 hallucinogens as part of their religious practices [17]. Recent discoveries indicate that the Mayans took intoxicating enemas using mixtures of these drugs. Many of these herbal extracts would normally cause gastric and intestinal upset, but they quickly gain access to the bloodstream with none of these problems when they enter during an enema. It is curious that a society that failed to use the wheel as a tool discovered instead an obscure aspect of physiology and generated an active syringe technology in order to give intoxicating enemas.

The GI tract is an interface operating to bring molecules into the body. The kidney, however, forms an interface operating to bring molecules out of the body. To do this, the kidney exploits a powerful combination of anatomy and physics to generate its physiology.

THE KIDNEY

The anatomy of the kidney is complicated enough that our first problem is to locate the interface we are interested in. Because the kidney does more than just filter things out, the interface is not so obvious. Additionally, the interface is not a single interface, but about two million individual interfaces that work separately yet function as one. Each kidney has about one million *nephrons*, which are the basic working units of the kidneys. Each nephron forms a complicated interface that combines with all the other interfaces to remove unwanted substances from the body.

So many nephrons give the kidneys some redundancy. An individual with only one kidney, for whatever reason, can get along quite nicely. But the additional duty placed on a single kidney causes it to grow or hypertrophy, increasing its size by increasing the size of the nephrons, but not their number. To unlock the secrets of the nephron, we must begin with the anatomy of the kidney. It is the anatomy combined with the powers of living cells that makes the kidney so effective.

Figure 5–6 shows the overall organization of a kidney and the anatomical reference points needed to follow the physiology. The outside portion of the kidney is called the *renal cortex*. This portion of the kidney contains the upper parts of the nephron. The next layer in is called the *outer medulla*. It contains the middle and lower segments of the nephron that are central to the concentration of urine. The innermost layer of the kidney is called the *inner medulla*. The inner medulla is part of the renal pyramids, which open finally into the urine collection depot that empties into the ureters. The ureters carry urine to the bladder, where it is stored until it is excreted through the urethra to the outside.

To position the kidney's function, we need to trace the outside surface of the body inward along the urethra and bladder, up the ureter, through the collecting ducts and into the nephron. Once the kidney forms a renal filtrate, this solution passes over the renal interface until final collection in the bladder. As the original filtrate passes over the interface, it is concentrated into its final form. The convolutions of the openings into the body can confuse our orientation, but the functional organization shown in Figure 5–1 will help us keep inside and outside correctly identified.

A portion of the blood supply must pass through the capillaries within the glomerulus. The glomerulus includes the primary renal capillaries and the glomerular space (see Figure 5–7). Here the first filtration process occur as blood pressure within the capillaries pushes the blood serum through the capillary walls into the glomerular space, while keeping the red cells and large proteins within the capillaries.

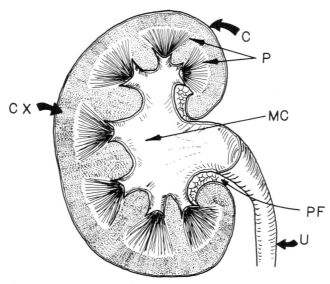

Figure 5.6. *Overall organization of the kidney. One of the busiest interfaces in the body is the kidney. A strong renal capsule (C) covers the kidney. The early filtration occurs in the renal cortex (CX). Urine collects in the renal pyramids (P) and empties into the major calyces (MC), draining finally into the ureter (U), eventually into the urinary bladder. PF is perirenal fat.*

But the rest of the nephron must receive blood, not only to supply nutrition and oxygen to the cells that carry on the kidney's processes, but also to carry back to the body those molecules and water conserved by this organ. The vascular network sends capillaries all along the nephron, extending deeply into the renal medulla. We will learn more about the role of these capillaries later.

The nephron is composed of six elements that not only have different functions, but also have different cellular organizations to carry out these functions. The nephron's anatomy begins at the *glomerulus* (Figure 5–7). Here blood vessels form a fine capillary lacework in intimate contact with the cells that form the glomerular capsule. A fine mesh of connective tissue separates the two spaces, vascular and glomerular, and in the mesh are small holes that permit the passage of fluids and small molecules. The red cells and many of the larger proteins are filtered out and remain in the vascular space. Following the glomerular space is the *proximal convoluted tubule*.

The urine then enters a large loop named for the man who discovered it. The loop has two identical parts, the *descending* and *ascending loops of Henle*. The cells within this loop are quite different in appearance and function from the remaining cells in the nephron. The ascending loop connects with the distal convoluted tubule, which

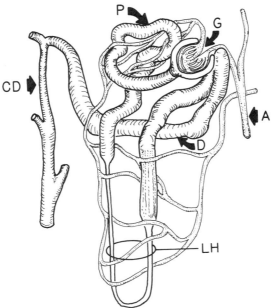

Figure 5.7. *Vascular and tubule organization of the nephron. The functional unit of the kidney is the nephron. The primary filtration occurs in the glomerulus (G), then urine formation begins in the proximal convoluted tubule (P), continues through the Loop of Henle (LH), and the distal convoluted tubule (D). The final step of urine formation occurs in the collecting duct (CD). An elaborate vascular network interlaces with the nephron tubules.*

in turn connects with the last structure of the nephron, the collecting duct. The two limbs of the loop of Henle and the collecting duct penetrate to or through the outer and inner renal medulla. This physical placement is instrumental in the operation of the nephron.

We can follow the function of the nephron by tracing events in a schematic nephron shown in Figure 5–8.

The function of the nephron begins in the glomerulus. Here the blood is filtered in a purely mechanical way. Driving the process is the blood pressure. The blood serum is simply passed through a sieve of sorts, and red cells and large proteins are filtered out to remain in the bloodstream, as the filtrate is brought into the glomerular space. At this point, the osmotic pressure of the filtrate is the same as that of the blood. Because the pores in the glomerular membrane are so small, the filtrate is nearly protein-free, containing only very small molecules that are dissolved in the blood serum.

The filtrate then enters the proximal convoluted tubule. Here the sodium in the filtrate is actively pumped out of the lumen back into

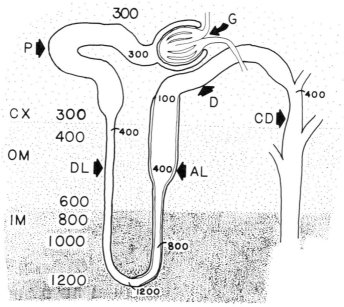

Figure 5.8. *The nephron and the osmotic gradient in the kidney. Primary filtration occurs in the glomerulus (G). The filtrate then passes along the proximal convoluted tubule (P) to the descending loop of Henle (DL). The urine then travels up the ascending loop of Henle (AL), which is impermeable to water, into the distal convoluted tubule (D), then into the collecting duct (CD). As the loop passes into the kidney from the renal cortex (CX) to the outer medulla (OM) and the inner medulla (IM), the tissue increases is osmotic pressure from 300 milliosmols to 1200 milliosmols. The numbers along the nephron represent the osmolarity of the urine at each point.*

the body. With this pumping process, using cell membrane-bound pumps like those found in nerve and muscle, goes the water. Thus, both sodium and water are removed back into the body in the proximal tubule. Near the midway point through the proximal tubule, about 67 percent of the original filtrate is back in the body. By the end of the proximal tubule, only 20 percent of the original filtrate remains. Now a very novel process begins that will concentrate the urine even further with very little energy expended by the kidney cells.

From the proximal tubule, the descending loop of Henle penetrates into the renal medulla. In the medulla, an increasing osmotic pressure exists as the loop progresses toward the inner medulla. The source of this osmotic gradient in the renal tissue is, in fact, in the loop of Henle. We will look at this formation later, but for now, we need to look at the effect of an osmotic gradient on the urine passing down the loop of Henle.

The walls of the loop are permeable to water and sodium, so as the urine moves toward the medulla, sodium diffuses into the urine from the tissue, and the water diffuses out, concentrating the urine in the process. At the bottom of the loop, the urine is only 15 percent of the original filtrate.

From the descending loop, the urine passes into the ascending loop of Henle, and here the cellular properties change. The ascending loop is *not* permeable to water, so despite the high concentration of sodium in the surrounding tissue, water cannot move out of the filtrate along the ascending loop. The ascending loop does, however, have a strong membrane pump for sodium, and sodium is pumped out of the urine as the urine moves toward the cortex and progressively lower concentrations of sodium in the surrounding tissue. Pumping sodium out of the ascending limb sets up the concentration of sodium in the surrounding tissue, making available the sodium that diffuses into the descending limb. The two limbs of the loop of Henle, ascending and descending, interact to form two currents counter to one another, producing the name "counter current mechanism" to describe this urine concentrating process.

Essential to the operation is the impermeability to water of the ascending limb. Once the osmotic gradient is set up, maintaining it requires very little energy. It is the combination of cell function and loop organization that makes setting up the gradient possible in the first place. Pumping sodium out of the urine works so well on the way up the ascending portion of the loop that urine is actually hypotonic (having an osmotic pressure less than blood) with respect to blood at the top of the ascending loop of Henle, and only 15 percent of the original filtrate remains.

From the ascending loop of Henle, the filtrate goes into the distal convoluted tubule. In this tubule, sodium pumps are once more at work, removing sodium from the urine. The water then diffuses out into a more osmotic tissue. At the end of this tubule, only 5 percent of the original filtrate is left, and the urine is once more isotonic (having the same osmotic pressure as blood) with blood, but highly concentrated. At the end of the distal convoluted tubule, the urine enters the collecting ducts. Once more, cellular function and cellular appearance change.

The collecting duct passes through the osmotic gradient again, but this time the permeability of the walls of the duct to water is controlled by a hormone secreted by the pituitary gland. The hormone is called *antidiuretic hormone* or ADH. This hormone makes the walls of the collecting duct become more permeable to water. As the urine passes by the hyperosmotic tissue, it loses water and becomes highly concentrated, about four times more concentrated than blood as the water follows the osmotic pressure in the medulla.

When ADH levels decrease, the collecting duct walls become less permeable to water, and more water stays within the duct, forming a more dilute urine, removing more body water in the process. Drugs that affect the secretion of ADH affect the ability of the kidney to form a concentrated urine. One such agent is *ethanol*, the alcohol found in liquor and beer. Ethanol suppresses the secretion of ADH, causing a more dilute urine than normal. As a result, a cold beer on a hot day can cause a net fluid loss from the body.

At the end of the collecting duct, only about 0.5 percent of original filtrate is present, forming a highly concentrated urine, getting rid of unwanted molecules, and at the same time conserving essential body sodium and water.

These concentrating features are possible because of the counter current formation and the unique capability of the cells in the nephron. Only humans and birds can form a highly concentrated urine, and only humans and birds have highly developed loops of Henle [21].

With these details in mind, let's review events in sequence. After the filtration in the glomerulus, the filtrate enters the proximal tubule, where sodium and water are removed to concentrate the urine. The filtrate then enters the descending loop of Henle, where sodium enters the solution while water leaves as the loop penetrates deeply into the highly osmotic medulla. At the bottom of the loop, the urine is about four times more osmotic than blood. Now with a great deal of water osmotically squeezed out, the sodium is once more removed by pumping, while the remaining water is kept within the ascending limb of the loop. At the top of the loop, the pump has made the solution hypotonic with respect to blood as it enters the distal convoluted tubule. In the distal tubule, more sodium and water are removed, concentrating the solution even more, but making it isotonic with blood. Now most of the unwanted molecules supply the osmotic pressure within the solution. The urine passes once more through the osmotic gradient as it flows through the collecting duct. Here ADH changes the water permeability of the duct wall, controlling the final concentration of the urine. In the whole process, the kidney forms a highly concentrated urine, recovering nearly 99.5 percent of the original filtrate.

The kidneys are quite busy forming about 160 liters of filtrate daily. That means that the total blood volume of a 150-pound human is filtered about 33 times every 24 hours.

Along with the removal of unwanted molecules, some of the body's essential molecules also appear in the filtrate, and they are recovered. So, in addition to the absorption of sodium and water, the kidney saves other compounds and ions like glucose. Too much glucose passing into the kidney can saturate its absorption mechanism, and sugar appears in the urine. This is the mechanism at work when sugar appears in the urine of an uncorrected diabetic. If more osmotic materials, glucose in this case, appear in the urine and cannot be removed, they cause an osmotic diuresis (an increase in urine formation because of the increased osmotic pressure within the urine), compromising the ability of the kidney to form a concentrated urine.

Ions like sodium and potassium are handled by the kidneys as well, contributing to the mineral balance of the body. In addition, the proximal tubule secretes acid into the lumen as the kidney contributes to the acid-base balance of the body. Acid secretion produces some other results. For example, penicillin is an acid in the body, and the kidney actively secretes it into the urine. In about four hours, 50 percent of the penicillin placed in the body is secreted and thereby removed from the circulation.

This rapid appearance of penicillin in the urine was utilized in the early use of this antibotic when it was still quite expensive and rare. Hospitals collected the urine of patients on penicillin and separated out the penicillin for reuse. The supply of penicillin is no longer a problem with mass production, and urine collection schemes are a part of history.

While a correctly functioning set of kidneys generates no special attention and is almost taken for granted, failing kidneys are another matter. People who have kidney failure suffer a lingering death unless some sort of filtration and blood cleansing system is set up. The modern kidney dialysis machine does this, taking over for the sick or absent kidney, filtering out harmful molecules. Kidney function is so essential to the body that the kidney is endowed with its own blood-pressure-regulating system that has priority over other blood-pressure-regulating systems.

Kidney function depends upon several mechanical factors such as blood pressure, blood flow, and urine flow. Diluting the osmotic gradient of the medulla can reduce the kidney's ability to concentrate urine. A profuse lymphatic system laces through the renal tissue, preventing any buildup of fluid within the tissue surrounding the tubules. Transplanted kidneys do not have the lymphatic system connected, and they work at a reduced efficiency as a result. Tissue rejection can cause inflammation and swelling in the transplanted kidney, and following the ability to concentrate urine is often a useful indicator of how much rejection is occurring in the transplanted kidney.

Because the kidney has a strong influence on the amount of extracellular water in the body, including the blood volume, drugs that affect renal function are used to control the blood volume, and thereby systemic blood pressure. To regulate its own blood pressure, the kidney uses a vasoactive hormone to control the systemic blood pressure.

A reduced blood flow to even a small portion of the kidney can increase the renin–angiotensin system activity, causing a serious elevation of blood pressure.

Although we have not explored all of the events that take place inside the kidney, its role in primary functions is clear. The kidney is an unusual organ, not for what it forms, but for how it carries out the formation. Urine is not a simple waste material. It is a physiological "success" formed by one of the most complex and powerful organs in the body. With understanding, urine will never again appear quite the same.

SUMMARY

Getting things into and out of the body is part of keeping the body alive. Several interfaces are sites of exchange for both molecular and nonmolecular entities. The skin, the lungs, the GI tract, and the kidneys are places where molecules are kept out, exchanged, absorbed, and removed from the body. A close examination shows that what we may have believed to be inside the body is not. And in this fine division between the inside and outside of the body, discrete lines of protection and function are set out.

The skin, with an area of about 1 to 3 m^2, is covered with several layers of dead cells that provide a watertight covering, preventing the loss of body water to the outside, and keeping the outside water from getting in. Under certain conditions, the skin does secrete water for cooling purposes. Covering the skin is a population of bacteria that are not threatening to the body because of the impenetrability of the skin. The skin, because of its strategic position, also transfers sensory information through nerve endings that reach close to the body's surface. Thus, the skin supplies several protective functions and is a site of nonmolecular exchanges with the outside world.

Unlike the skin, the lungs are the primary site for gas exchange, namely, oxygen into the body and carbon dioxide out of the body. They have a functional area of about 70 m^2. The actual sites for gas exchange are the alveoli, which are globular sacs at the ends of the multibranching airways. Because of the blind-sac design, the lungs must expand physically to bring in air, and collapse to expel air, with each breathing cycle bringing in oxygen and expelling carbon dioxide and water. The lung interface appears thin and fragile, seemingly open to easy infection and diseases produced by the materials we breathe, yet its organization gives it strength and adaptability.

In contrast to the lung, the GI tract seems less fragile. In these structures, food is chewed, digested, and passed by an interface, while molecular transfer is made into the bloodstream. Molecules that enter the body through the GI tract must withstand the stomach acid and enzymes to finally make it into the body. Three segments of the small intestine provide most of the absorption, and the colon or large intestine provides final dehydration and compaction of the food residue. Because the colon is so well vascularized, it is a good site for getting agents into the body without nausea, vomiting, or diarrhea.

Unlike the GI tract, the kidney turns out to be an anatomically complex organ, filtering the circulation and removing water-soluble molecules from the body. Exploiting a unique anatomical relationship and selective powers of living cells, the kidney is able to produce a urine about four times more concentrated than blood. Although the channel is convoluted and passes through several other organs on the way to the outside world, urine, after its initial filtration, is effectively outside the body. Urine concentration and molecular removal comes about because of selective secretion into and absorption from the lumen of the nephron. In the end, the kidney is a small organ utilizing powerful physics and chemistry.

Getting things into and out of the body is not easy. Each interface applies special qualities that keep the hypodermic needle part of current events rather than ancient history. New technologies may come up, such as innoculating air guns or special carrier molecules like DMSO, but all are attempts at getting around the special qualities of the body's interfaces with the outside world.

6

CONCEPTION TO BIRTH: A MOMENTOUS JOURNEY

We have all been in the uterus, although being there is not part of conscious memory, and it takes some imagination for us to think about the fetal experience. It is warm in the uterus, and there is probably little feeling of weight because of the suspension in the amniotic fluid. It is probably dark most of the time, with some sounds penetrating to the fetal ears, including those of the mother's voice and her beating heart. The sounds that do penetrate from the outside are probably distorted by the intervening fluid and are similar to what one would hear submerged in a swimming pool. Physicians routinely use the fetal response to sounds to measure a fetal head diameter. Using ultrasound (a form of imaging using very high frequency sound above human hearing), they measure the diameter of the fetal head to determine the stage of growth by tapping on and stroking the mother's abdomen to make sounds that cause the fetus to move for the best image of the fetal head. Newborns show a clear ability to recognize and respond to their mother's voice [1]. The fetus can hear and learn in the uterus.

Few things have opened the world of the fetus to medicine as much as real-time, ultrasonic imaging. Until the arrival of ultrasound, we had only brief glimpses into the uterus with X rays and endoscopes, all putting the mother and child at risk. Ultrasound changed that, giving us opportunities to see a living fetus go about its normal activities. Along with an image of a 12-week-old fetus having hiccups or sucking its thumb comes the essential information that can help an obstetrician detect and treat a pregnancy that is veering away from normal.

188

The journey that begins with a single fused cell and ends with a developed, functioning human, capable of independent life, is filled with constant change. Along the way, a single cell will divide an astronomical number of times, invade the body tissues of the maternal organism, and remain undetected by the maternal immune system. In a span of 40 weeks, growth and cellular specialization will complete an organism that will undergo yet another major metamorphosis at delivery, moving from one physiological world into another. The fetus will change from a nutritional parasite into a small, physiologically independent human, capable of breathing air and eating food. In the change from uterus to "outside," the fetus will reorganize its vascular anatomy and physiology in a major way, and it will do so in a very short period of time or not survive. Early on, these changes are stable, but not physically complete. With time, the fetal ducts will disappear into a scar and a connective tissue ligament, a common course of events for most people. Not everyone completes the transition, however. For a small portion of the adult population, the *foramen ovale* remains functionally but not anatomically closed in a small but significant number of people. Birth is a time of change and adaptation, and a time of risk, when survival itself is sometimes in question.

The two greatest periods of change for the fetus occur at both ends of the gestational process. The first period starts at the formation of the fertilized egg, termed a *zygote*, which then becomes an *embryo*, and ends when the embryo becomes a *fetus*, which occurs near the end of the eighth week of development when the embryo has all its rudimentary organs. The second period is at birth, when the fetus becomes a newborn or *neonate*. Not surprisingly, these are also the two periods when the *conceptus* (which includes both the developing human and the placenta) is most vulnerable to outside forces that could injure or kill the new life. The journey may seem perilous, but it is successful for most of those who undertake it, punctuated by some of the most astonishing events we call developmental biology.

CONCEPTION: UNITY AND EARLY PLANS

One of the more difficult aspects of reproduction to grasp is the connection between a single cell, so small that it sits at the resolution limit of the unaided eye, and the newborn still curled into a fetal position in a hospital nursery. From that single first cell came the instructions for all the later daughter cells to divide, specialize, and gather together to form organs. All this information comes from the genetic information coded on the cellular DNA. The genetic "dose" for this cell comes from the maternal ovum and the paternal sperm, each normally contributing half of the genetic complement. The fertilized cell may be small, but its information is complete.

The cells formed for reproduction go through a division different from normal mitosis. The process, called *meiosis*, produces a cell with half the original number of chromosomes. For humans, it means a final nucleus containing 23 chromosomes. Reaching the final cellular form requires two divisions, however, one to split the chromosomes into daughter cells and a second to replicate the daughter cells. Thus, from a single germinal cell comes two daughter cells that are *haploid* (one half the chromosomal complement), which divide again without changing the chromosomal number to produce the final cells used in reproduction.

But the divisional process is not identical for both male and female. For a male, meiosis means that a single germinal cell will produce four sperm cells. For a female, the process differs slightly, producing only one ovum and a so-called *polar body* at the first division of the original germinal cell. The second division awaits completion at fertilization. The two polar bodies are ultimately dissolved, leaving only one active set of genes for fertilization.

The effective lifetime of sperm and ovum is also different. Spermatogenesis is a life-long activity for a normal male, with a constant production and reabsorption in the testes that continually renews the available sperm cells. The ova, however, are not renewed throughout a woman's life, and an ovum released in her 30th year is a cell 30 years old. And the process that selects which ovum out of the many available will be released for any menstrual cycle is yet to be understood.

The ovum is usually released about midway through the female reproductive or *estrus* cycle. Surrounding the released ovum are other cells and barriers forming the *zona pellucida* next to the cell membrane, and an outer collection of cells called the *corona radiata*, made of adhering follicular cells from ovary. On the ovary that released the ovum remains the *corpus luteum*, which becomes a source of the hormone *progesterone* during its active phase.

A released ovum enters one of the *Fallopian tubes* and travels down the tube toward the body of the uterus, propelled by the Fallopian tube (Figure 6–1). Fertilization usually occurs in the Fallopian tube when the ovum is met by one or more sperm cells traveling up the Fallopian tube from the uterine cavity (Figure 6–1). The relatively long distance the sperm must travel from its deposition site in the vagina to the site of conception sets up a physical selection process, demanding that only the more vigorous sperm reach the ovum to fertilize it.

In addition, the sperm is conditioned for fertilization along the way. The conditioning process seems to remove a covering on the sperm that would prevent its activity with the ovum. A second process

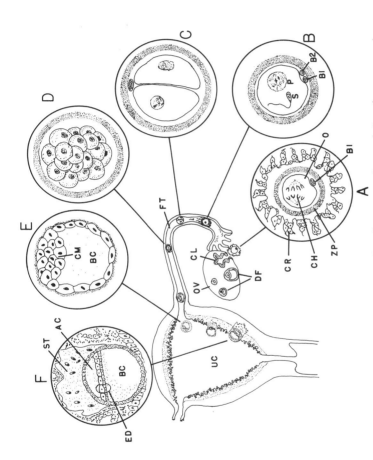

Figure 6.1. *The early events in fertilization and implantation. Before ovulation, the ovary carries several developing follicles (DV). The mature follicle bursts, releasing an ovum into the fallopian tube (FT). The remaining follicle tissue on the ovary forms the corpus luteum (CL). A. The ovum has an outer coat of cells called the corona radiata (CR), then another layer called the zona pallucida (ZP). The ovum (O) has chromosomes (CH), with a polar body nearby (B1). B. Fertilization with a sperm (S) causes the ovum to complete meiosis forming a second polar body (B2), then an ovary pronucleus (P). The sperm will form a similar structure before chromosomal fusion. C. The two-cell stage of division. D. Formation of morula, multicell cluster. E. Formation of a blastocyst with an inner cell mass (CM) at one pole. F. Implantation involves an invasion of uterine tissue by the syncytiotrophoblast (ST). The blastocyst forms the embryonic disc (ED), separating the amniotic cavity (AC) from the blastocystic cavity (BC). UC is the uterine cavity.*

191

called *capitation* changes the head or *acrosome* of the sperm, exposing enzymes the sperm can use to penetrate to the mature ovum through its protective barriers [2].

The first sperm to make its way through the layers of the ovum will penetrate through the ovum's cell membrane. At this union of sperm and ovum, the ovum becomes impermeable to all of the remaining sperm, ensuring that only one sperm can join with the ovum.

At fertilization, the ovum completes its second meiotic division, forming the zygote and the second polar body. The first and second polar bodies will disappear, clearing the way for male and female DNA fusion and developmental mitosis to begin. At the completion of fertilization, the cell begins to divide quickly as it continues its journey down the Fallopian tube.

In general, each cell must have only one X sex chromosome to function properly. A genetic male conception will normally have one X chromosome, contributed by the ovum, and a Y chromosome from the sperm. On the other hand, a genetic female conception will have two X chromosomes, one from the sperm, another from the ovum. At about the 8- to 16-cell stage, one of the female X chromosomes is placed in storage in each cell, forming a dark-staining region in the cell nucleus called the *Barr body* [3]. Counting Barr bodies in a cell also counts the number of X chromosomes in that cell. The presence of more than one Barr body indicates chromosomal disease.

As division continues, each daughter cell becomes smaller in size, with the cells forming a cluster that looks like a mulberry. At this stage, the cell cluster is called a *morula* (Figure 6–1).

As the morula enters the uterine cavity, it begins to take in fluid, forming a cyst with a small mass of cells at one edge. The structure is now called a *blastocyst*. The outer perimeter of the cell cluster is called the *trophoblast*, a layer of cells that are destined to have an important role in survival of the conceptus, a role that starts at this stage and ends at birth. The blastocyst attaches to the uterine *endometrium*, and the trophoblastic layer begins a cancer-like invasion of the uterine wall. The invasive tissue is a syncitium, a fusion of many individual cells, called the *syncytiotrophoblast*, and looks like a single, giant, odd-shaped cell with many nuclei.

Meanwhile, the blastocyst is also undergoing an organizational change, forming a disc of tissue with fluid on both sides. One fluid collection is the *amniotic cavity*; the other is the *blastocyst cavity*. The disc of cells is called the *embryonic plate* and marks not only an anatomical stage in development but also identifiable sources of the different tissues in the mature body [2]. The conceptus is now an embryo (Figure 6–1).

From here on, the embryo will differentiate cells into tissues that will form organs and structures for the developing body. Where do the instructions for these changes come from? How can cells sitting side by side develop into entirely different organs with entirely different functions? Answering these questions has been the major task of researchers working in developmental biology. And some of the answers are astonishing.

As the cells divide, even at the first cell division, dedication and programming are already occurring. The daughter cells do not receive an equal distribution of cell materials, either in quality or quantity. Tracing the instruction back, investigators have found that even the unfertilized ovum is not homogeneous [4]. Unevenly distributed within the cytoplasm of the ovum is information that dedicates the cell's progeny. The message-carrying substances turn out to be messenger-RNAs replicated from the ovum's DNA [4]. The RNAs are *not* evenly distributed in the cell, and cell division is along specific anatomical lines that unevenly direct the RNAs into the daughter cells. The source of this cytoplasmic organization seems to be the microfilaments and microfibrils now known to give the cytoplasm organization and structure [5].

The RNAs, for the most part, seem to encode the production of *histones*, proteins found closely allied with DNA. Microscopically, the histones combine with DNA to form "beads" on the DNA strands with pure DNA strung out between the beads [4]. In addition, the histones appear to turn genes off rather than on. As a result, it seems that differentiation comes from a progressive, selective inhibition of genes rather than the production of "promotors" that turn them on.

At implantation into the uterine wall, a new set of problems arises. The fetus has a genetically individual tissue, and implantation places the conceptus directly in contact with maternal tissue and an intact maternal immune system.

GESTATION:
AVOIDING MATERNAL REJECTION

To understand the unique position of the embryo and fetus in the scheme of things, we need to understand a few things about events surrounding the grafting of tissue from one individual onto another.

We can take a piece of skin from one individual and graft it into place on another individual. If both individuals are of the same species, the graft is called an *allograft* or a *homograft* [6]. In about one to two weeks after the grafting procedure, the graft begins to undergo a series of changes that lead to a rejection of the donor tissue by the host. The graft tissue dies and sloughs off.

Rejection comes about because the host organism recognizes the donor tissue as "foreign." This recognition is possible because each cell carries "identity molecules" on its surface that tell the organism that these cells belong to itself. Thus, an organism made of many cells sets up a recognition system of "self" and "not self." Cells that do not belong to the organism stimulate a response in the host to locate all the foreign cells and remove them from the host.

The host's rejection process starts when a mobile lymphocyte recognizes one of the graft cells as "not self." Recognition ultimately stimulates the lymphatic tissue into creating sensitized lymphocytes or *effector cells* [6]. They invade the foreign tissue, attacking and destroying the donor cells. The lymphocytes are sensitized to the "not self" flags on the graft and attack only that particular tissue. Should a second graft from a third individual be placed on the host, the effector cells working on graft one will not recognize graft two. The second graft will later receive its own set of sensitized effector cells.

An immune response can involve more than the effector cells, however. Another response can come through protein-based antibodies that are structure-specific [6]. The antibodies combine with enzymes to form *complement*, an immunological substance that works to destroy donor lymphoid– and blood-forming tissues [6]. The antibodies are formed to attach to the stimulating *antigen*—that is, the material the body interpretes as foreign.

When the embryo implants into the uterine tissue, it becomes a classical allograft: a genetically foreign tissue in contact with a host tissue. And the conceptus remains an allograft throughout gestation. Why is it not rejected? Finding the answer led not only to a better understanding of the mechanisms of gestation but also pointed the way to understanding some unusual neonatal diseases. Let's initially look at some logical sources for this protection for the fetus.

First, the growing fetus *in utero* is genetically different from the maternal tissue but shares with it a common genetic heritage. Grafting fetal tissue from outside the uterus onto maternal tissue leads to a standard rejection sequence; thus it is apparent that the common genetic history of mother and fetus affords no special protection [6]. Neither does the fact that the tissue happens to be fetal.

Second, the uterus might afford protection. Allografts onto the uterine endometrium with a hormonal environment that looks like implantation and pregnancy still experiences uniform rejection [6]. Neither the uterus nor the hormones of pregnancy provide special protection for the fetus.

The sequence of implantation and the specific interfacing tissues between the two systems point to the tissue responsible for this fetal protection. The tissue surrounding the conceptus that invades the endometrium is the *trophoblast*, with the *cellular trophoblast* laying down an insulating layer of cells between fetal and maternal circulatory patterns [7]. Thus, despite the close proximity of the two blood systems, they appear physically and immunologically separated. The cellular trophoblast maintains this separation.

The separation is not total, however. This is evident when the fetus carries a major antigen lacking in the maternal system. The so-called Rhesus or Rh factor in blood and the ensuing maternal reaction provide a good example. In a situation where the fetus carries the Rh factor (Rh positive) and the mother does not (Rh negative), the maternal immune system can form humoral antibodies against the fetal Rh factor [8]. These humoral antibodies can cross the placenta, causing *erythroblastosis fetalis*, a large-scale destruction of fetal red blood cells, leading to a lethal form of fetal anemia. Therapy is a complete replacement of the fetal blood volume, removing both the damaged fetal red blood cells and the active humoral antibodies.

The maternal immune system becomes aware of the Rh factor because the two systems, fetal and maternal, are not absolutely isolated. In fact, the two systems exchange mobile cells on a regular basis, which are usually leukocytes and lymphocytes [6]. The exchange is not large, but it is still often large enough to trigger some immune response in the maternal immune system, leading to lymph-node hypertrophy in the lymphatic system draining the uterus [6]. Still, for most pregnancies, the fetal protection remains intact as the cellular trophoblast inactivates the few maternal lymphocytes that breach the separation.

Under normal conditions, the trophoblast handles the migrating maternal lymphocytes, coating them with antibodies of its own that keep the maternal lymphocyte harmless to the fetus [6]. This property of the trophoblast can be transmitted elsewhere on the body. Allografts using trophoblastic tissue are not rejected by the host [6]. Thus, slipping by the potential maternal tissue rejection requires an intact, healthy, fetal cellular trophoblast.

If too many maternal lymphocytes make it through the placenta, however, and swamp the trophoblastic protection, then active, sensitized lymphocytes can enter the fetus. This produces a graft-versus-host disease in the fetus, where the invading maternal lymphocytes can depress the normal fetal growth pattern, producing a "runt" who fails to thrive and dies soon after birth [6].

As implantation and development proceed, more conventional biological controls begin to take over, namely, hormones, and they show up in the amniotic fluid.

HORMONES *IN UTERO:* ARCHITECTS OF CHANGE

Beyond eight weeks' gestation, the fetus has all its organs, but they are at various stages of development. Yet the fetus must begin to work as a complete system, exercising fetal control over the developing organs and structures. As we learned earlier, hormones are one of the instruments controlling body functions, and as hormonal control becomes more complete, many of the hormones, their precursors, and metabolic byproducts appear in the amniotic fluid.

From this point of view, the amniotic fluid is not just a large fluid collection to absorb physical shock. It is, instead, a flowing, changing environment that is often a route for fetal clearance of unwanted substances and intermediate compounds that appear in the biochemistry of the conceptus. Examining the hormonal content of the amniotic fluid provides clues to some of the events in the development of fetal biochemical maturity.

The hormones can also be more than a simple step in fetal development. They can provide a glimpse of the mutual support between the conceptus and the maternal system to maintain pregnancy. The protein hormones include: human chorionic gonadotropin, human chorionic somatomammotropin, prolactin, glucagon, and human growth hormone. Steroid hormones include: C16 (Carbon location 16) hydroxylated steroids, estrogens, cortisol, progesterone, pregnanediol, and androgens. Prostaglandins of the types E, E_2, F, and $F_{2\ alpha}$ also appear in the amniotic fluid. A brief look at each shows how much remains to be discovered about the complex process of pregnancy and birth.

Human Chorionic Gonadotropin

Human chorionic gonadotropin (HCG) is produced and secreted by the trophoblast into the maternal circulation [9]. The secretion favors the maternal circulation with the greatest concentration there, then the amniotic fluid, with the lowest concentration in the fetal circulation. HCG shows a marked rise in concentration at the ninth week, reaching a maximum at about the thirteenth week, then decreases steadily to a low level at the fortieth week.

HCG seems to be connected with testosterone secretion in the male fetus, stimulating the descent of the fetal testes into the scrotum [10]. Because HCG appears in relatively large quantities early in the pregnancy, it is central to some chemical pregnancy-testing systems.

Human Chorionic Somatomammotropin

Human chorionic somatomammotropin, or HCS (also known as human placental lactogen), like HCG, is secreted primarily into the maternal circulation [9]. It begins a rise in the maternal circulation at 11 to 13 weeks that continues to the fortieth week. In contrast, the amniotic fluid concentration decreases until it is significantly less than one-tenth the maternal level.

The route of HCS into the amniotic fluid is unknown, and the placenta does not show a preferred pumping direction for this hormone [9]. The sources of the concentration imbalance may be an increasing metabolism in the amniotic fluid or a decreasing transfer of the substance through the amniotic membrane in the placenta as the pregnancy progresses.

HCS has been found to reduce the secretion of maternal growth hormone and may function as a special form of maternal growth hormone during pregnancy [11]. In addition, the secretion of HCS tracks with the size of the placenta; thus, lower-than-normal levels of HCS in the maternal circulation can indicate problems with placental development [9].

Prolactin

Amniotic levels of prolactin are very high early in gestation, nearly 100 times the maternal concentrations [9]. As the gestation progresses, this hormone decreases its amniotic concentration, ending up close to the normal maternal levels.

Every comparison of amniotic prolactin with pituitary prolactin shows that they are identical [9]. The source of this hormone is thought to be the amnion, but definitive evidence is lacking. Amniotic prolactin seems to be a major factor in development of the maternal mammary glands. In addition, the hormone inhibits the insulin-mediated active transport of glucose into cells [12]. Its role in fetal development that would require an increased amniotic concentration is not known, but some evidence suggests a contribution to amniotic fluid osmolarity and volume control [9].

Glucagon

Glucagon concentration in the amniotic fluid is nearly the same as that in the maternal and the fetal circulation [9]. Both the chorion and amnion are impermeable to glucagon, however. The source of glucagon in the amniotic fluid is not currently known.

Glucagon in the adult is a functional antagonist to insulin, working through the liver to increase blood sugar as insulin decreases it. Although glucagon may have the same role in the fetus, details on its position in fetal metabolism are not known.

Growth Hormone

On the maternal side, growth hormone is a protein hormone secreted from the maternal pituitary gland. Throughout pregnancy, the maternal levels of growth hormone are constant. In contrast, the amniotic fluid levels of growth hormone slowly increase throughout gestation [9]. Thus, the logical source of the amniotic growth hormone is the fetus, and fetal serum measurements show its concentration to be higher than that of the amniotic fluid.

Hydroxylated Steroids

In normal metabolism, inactivation of adrenal steroid compounds is through hydroxylation at the carbon 17 position of the steroid nucleus (based on the cholesterol molecule) [13]. The fetus, however, has an increased ability to inactivate steroids by hydroxylation at the carbon 16 position, which sets its liver function apart from expected metabolic pathways [9]. This seems to provide a means of fetal excretion of the steriods through the fetal urine into the amniotic fluid.

In some diseases, the fetal liver loses its ability to carry out hydroxylation at the carbon 16 position [9]. As a result, the concentration of non-C16 hydroxylated steroids increases, and this can indicate an altered fetal hepatic function.

Estrogens

At least three major estrogens are present in the amniotic fluid: estrone, estradiol, and estriol, along with modified forms of these estrogens as weak acids [9]. Amniotic fluid also carries six additional, less-well-studied estrogens in small amounts. Estrone and estradiol are at the end of two known synthetic pathways for steroids in the adrenal cortex [14].

Estrone and estradiol are rather steady from 12 weeks until birth. Estriol, however, increases slowly but steadily throughout pregnancy, reaching its highest concentration at birth after a sharp rise in concentration at about 33 weeks of gestation [9]. This pulse of estriol often results in an amniotic concentration four or five times higher than the initial concentration.

Producing estriol requires an intact feto-placental unit and synthesis from fetal and maternal precursors [9]. The placenta is the primary source of estriol for the fetal environment, with another form of estriol coming from the fetus.

A sudden decrease in estriol levels in the amniotic fluid seems to be correlated with a serious Rh-factor immunization and cases of abnormal fluid accumulation in the uterus (hydramnios). About 80 percent of the amniotic estriol seems to come from the fetus. As a result, estriol is often used to estimate chemically the fetal well-being.

Cortisol

Fetal adrenal morphology is different from that of an adult, and the presence of amniotic cortisol indicates, at the outset, functioning fetal adrenal glands [9]. Along with cortisol is cortisone and several other 17-hydroxycorticosteroids (a special class of steroids) in the amniotic fluid.

The amniotic fluid concentrations of the corticoids correlate with the umbilical levels, but not maternal levels [9]. The conceptus seems to be the source of these compounds.

The corticoid concentrations rise in the second trimester of pregnancy and spiral upward just before spontaneous labor. In general, they are higher in spontaneous labor than in induced or caesarian sections without labor.

These hormones play a yet-to-be-determined role within fetal physiology and also have a major influence on fetal lung maturity at birth, which we will look at later in this chapter.

Progesterone and Pregnanediol

The placenta produces progesterone, and this hormone increases concentration in both the maternal and fetal serum throughout pregnancy [9]. Some progesterone secretes into the maternal circulaion, but most is used by the fetus as a precursor to many steroid molecules [9].

As pregnancy progresses, amniotic levels of progesterone decrease, but it still appears in both a water soluble and water insoluble form, suggesting sources other than fetal kidney excretion. Along with being a precursor to steroids of various type, it may also play a role in labor.

In adults, pregnanediol is an excretory form of progesterone, usually in a water soluble form that the kidneys can excrete [13]. This seems also to be the case for the fetus with water soluble forms of this hormone appearing in the amniotic fluid, presumably from fetal urination.

Androgens

The production of androgens in the uterus probably reflects fetal sex steroid production, but in some instances it indicates abnormalities in fetal steroid production. For example, it is found in high quantities when the fetus has congenital adrenal hyperplasia (a defective growth in the adrenal cortex) [9].

In general, androgens are found in higher concentrations when the fetus is male. At about 20 weeks' gestation, testosterone is higher with male fetuses than in female fetuses, but testosterone becomes a less reliable indicator of sex closer to birth [9].

Prostaglandins

Prostaglandins are a relatively new class of hormones, and their function in the uterus is still being studied. They may have a role in controlling gestation and labor [9].

With implantation and development of hormonal control, the fetus lives in a confined, specialized world, growing, changing, and preparing for birth.

BIRTH:
TRANSITION INTO A NEW WORLD

The Fetal Condition

Within the uterus, the odd-shaped fetus is suspended in amniotic fluid that has a chemical composition like that of sea water. It is a small sea, suspending and protecting the fetus, all bounded by the smooth muscle of the uterus. This fluid volume provides a protected environment for the growing human to test and develop skills before they are needed outside. The amniotic fluid is also a stable, clean environment to grow in. For example, as the fetal digestive tract develops and matures, the fetus regularly swallows quantities of amniotic fluid. This fluid passes into the fetus's body through GI absorption, the fetal kidneys then remove it forming urine, and the urine is then voided back into the amniotic fluid. To keep things in balance with all this fluid movement, rather sensitive regulators hold the amount and quality of the amniotic fluid stable over the gestational period.

The quality of the amniotic fluid also plays a role in the developing respiratory system. Diagnostic ultrasound shows frequent attempts by the fetus to "breathe," moving fluid into the lung passages by major chest movements [15]. Despite all this activity, the amniotic fluid remains relatively "clean," supporting life *in utero*. Yet how all this is managed in the uterus is still a mystery.

The source of nutrition and oxygen for the fetus is the *placenta*, a highly specialized organ that provides an intimate contact between fetal and maternal circulations. The architecture of the placenta lets maternal and fetal blood pass very close to each other, at times separated by no more than the thickness of a single cell, but the two blood-pools do not mix [16]. The placenta is the exchange site for oxygen, carbon dioxide, and all of the atoms, ions, and molecules used to build the fetus. At the same time, all the waste products of cell metabolism produced by a growing fetus are removed by this same placenta. The exchange area in the placenta is very busy, for all these molecular exchanges occur through an area one-half the size of the newborn's lung area [17]. Further, many of the diffusion distances are as much as seven times the neonatal lung thickness [18].

The disadvantages do not stop here, however. Fetal arterial–blood oxygen pressure in the placenta is not equilibrated against the maternal arterial oxygen pressure but against the much lower maternal venous oxygen pressure.

The arrangment of the placenta within the fetal circulation is shown in Figure 6–2. Even the organization of the placenta within the fetal circulation is unusual, for the placenta is not in series with the body circulation like the adult lung, but in parallel. Clearly, the fetus seems to have several impediments to getting oxygen.

Even with an unusual diffusion arrangement and flow pattern, it is a false notion to think of fetal circulation as abnormal. It is, when we compare it with the adult pattern, certainly different, but it is not abnormal. The adult circulatory pattern is as incompatible with intrauterine life as the fetal pattern is to extrauterine life. Each form is appropriate for living conditions within a particular environment.

What makes the fetal circulation compatible with life are the special vascular ducts, a parallel arrangement of the placenta with the remainder of the circulation, and a parallel arrangement of the heart's chambers. Tracing out the circulation pattern, beginning at the placenta, will illustrate the main features.

Blood, oxygenated in the placenta, mixes with venous blood coming from the fetal systemic circulation, lowering the final oxygen tension of the oxygenated blood in the process. The blood enters the heart through the inferior vena cava. Through the atrial wall that separates the right from the left atrium (the interatrial septum) is a large opening called the *foramen ovale*. This opening provides direct communication between the two atria, and as blood flows into the right atrium, the shape of the right atrium divides the moving blood between both the right and left atria. Because of the flow mechanics at the entrance of the *foramen ovale*, the division of blood flow favors the left heart [19]. On the left atrial side of the foramen is a flap valve, connected to the upper edge of the opening. This flap is made of a combination of muscle and connective tissue and is kept open by the kinetic energy of the blood flow channeled through the foramen into the left atrium [20]. Viewing the heart as a pump, the *foramen ovale* provides a parallel input to the right and left heart chambers.

On the left side of the pump, the *ductus arteriosus* brings together the pulmonary artery and the aorta, letting both sides of the heart empty into the systemic and placental circulations at the same time (Figure 6–2). The *ductus arteriosus* is a well-muscled channel with

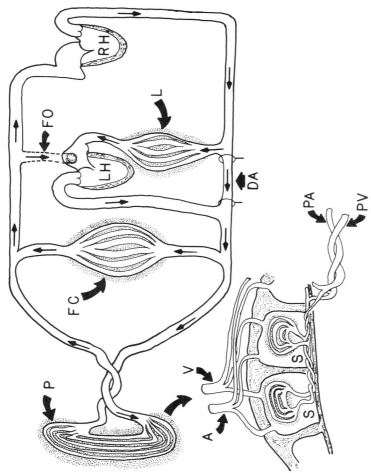

Figure 6.2. *Fetal circulatory pattern. The left heart (LH) and right heart (RH) are in parallel due to the ductus arteriosus (DA). The lungs (L) have only enough flow for tissue growth. The placental circulation (P) is in parallel with the fetal circulation (FC). Fetal (PA, PV) and maternal (A, V) circulations come in contact through the placental sinuses (S), where the exchange takes place. The cellular trophoblast keeps the two blood pools separate.*

both sensory and motor nerves [21]. The exact function of the nerve tissue is not known. The musculature, however, assures that at the right time, the duct will close as part of the circulatory pattern changes at birth.

Because the special ducts modify flow to such an extent, fetal lungs receive only a small portion of the overall blood flow. For example, measurements indicate that fetal lungs get about 10 to 15 percent of the total cardiac output [22]. Because the vascular requirements are only to keep the lung tissue alive and growing, this is an adequate amount of blood flow.

A look at the adult circulation (Figure 6–3) shows the uniqueness of the placental circulation. In the adult pattern, a single pass through the lungs oxygenates the blood, with the two sides of the heart in series. In contrast, fetal blood must divide between the placenta, where oxygenation occurs, and the systemic circulation, where oxygen is consumed. But because of some subtle fluid mechanics in the fetal vessels, about 75 percent of the blood pumped by both sides of the heart (total cardiac output) passes through the placenta rather than the systemic circulation, oxygenating a large portion of the fetal blood [23]. For its body size, the fetus possesses a rather high heart rate and sustains nearly three times the adult cardiac output. This high flow rate appears to compensate for the unusual fetal circulatory pattern, keeping the movement of blood through the placenta and fetal body adequate for normal growth and development.

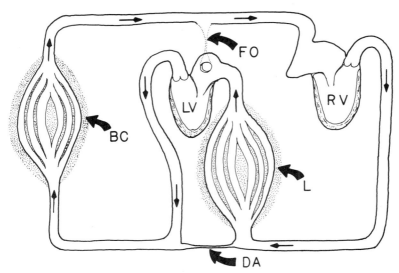

Figure 6.3. *The neonatal circulatory pattern. The two sides of the heart are now in series, and all cardiac output goes through the lungs (L). The fetal channels (foramen ovale, FO and ductus arteriosus, DA) are still present. LV is left heart, RV is right heart, and BC is body circulation.*

The relatively high fetal cardiac output also compensates for the unusual flow pattern in the placenta, which, as pointed out earlier, is set up for oxygen equilibration against maternal veins. The pattern, shown in Figure 6–2, is called a concurrent pattern in which fetal and maternal blood flow in the same direction during gas exchange.

Adding the equilibration against maternal venous oxygen pressure to the other impediments, the character of the impediments to fetal oxygenation appears impossible to overcome. And maternal behavior can add to these impediments. If the mother lives at a high elevation or is a smoker, the fetus is often born with a lower-than-normal body weight [24]. Combining high elevation and smoking can push the fetal weight down even lower.

The fetus, on the other hand, counters these forces that reduce tissue oxygen by using a special hemoglobin designed to have a higher affinity for oxygen than adult hemoglobin [25]. This hemoglobin, called *hemoglobin F*, can become about 13 percent more saturated with oxygen than adult blood at the same low oxygen pressures. This seems a small adaptation, but it is a very effective way of satisfying fetal oxygen needs.

Despite an organization that seems to slow fetal oxygen supplies, fetal tissue growth and energy consumption are astonishing. But the fetus needs more than oxygen; it needs amino acids and a long list of other molecules and minerals that make up the body and supply it with energy. And all these substances can reach the fetus only through the placenta.

Along with the nutritional requirements, the fetus must control its body temperature. The fetus can receive heat energy three ways: through the placenta, carried by mother's blood; through the amniotic fluid; and by fetal metabolism. The fetus seems to have a fairly constant temperature, but its small size and relatively large surface area make heat transfer to and from the fetus very easy. Normally, body temperature regulation is mediated through the hypothalamus. But like autonomic control of the cardiovascular system, the ability to control body temperature may not come until late in gestation, when the fetal nervous system is adequately developed. The increased tissue building may generate a lot of local heat, which moves to the maternal body through the placenta and amniotic fluid. The amniotic fluid, in addition to its other functions, may also provide a heat buffer while the fetal temperature control is still immature.

During this gestational period, the fetal lungs are collapsed, compressing the pulmonary vessels and giving the pulmonary vascular bed a high vascular resistance. Up to the first few breaths, the lungs are not functional tissue, receiving only a small supply of blood to build tissues and prepare for the task that begins at birth.

Many of the reflexes and activities that are normal for the neonate are practiced *in utero*. We know this from the ultrasonic images that show the fetus in its watery world. The fetus can be seen stretching, kicking, sucking, drinking, and testing its breathing apparatus by inhaling amniotic fluid. In the late portion of the third trimester, the fetal bladder fills with urine about every 30 minutes and empties into the amniotic fluid [26]. The fetus regularly ingests amniotic fluid and forms fetal feces called *myconium.* The cycle of fluid is complex, and all of the routes are not known.

With a continual circulation of amniotic fluid to provide a safe, clean environment, development continues, fueled by oxygen and nutrients flowing through the placenta, cued by hormones secreted by fetal tissues, fetal and maternal glands, and the placenta. If we could hear hormones, the constant communication among cells and tissues might sound like the floor of a stock exchange. With everyone shouting and passing messages, it all appears to be pandemonium, yet a sense of order underlies it all. So it is with the cells that ultimately give the fetus life.

As intrauterine growth nears its finish, the fetus changes position, moving head down, preparing for delivery.

Entering the New World

While the fetus was growing and developing skills it will use to survive in the new world, the uterine contractions that began as only occasional slight tightenings of the uterine smooth muscle walls now come closer together and become more intense as birth approaches. The rhythm of the contractions is set by a group of cells called pacemaker cells. Just how the rhythm in these cells is set remains obscure, but the rhythm increases in frequency and intensity with approaching birth.

The fetal head pushing and wedging into the uterine opening stimulates uterine contractions. Helping things along is a hormone called *oxytocin,* secreted by the maternal anterior pituitary, which increases the intensity and frequency of the uterine contractions. At the same time, the uterus becomes unusually sensitive to this hormone. The uterine contractions feed back through the nervous system to stimulate further release of oxytocin. This interaction of uterus and pituitary gland form a positive feedback system (see Chapter 4 on controllers).

All these events occur near the thirty-ninth to fortieth week of gestation, and like the heart, which increases both its intensity and frequency of contractions when stretched, the uterus contracts more strongly and frequently, responding to the stretch from the fetal head, helped by the increasing levels of oxytocin.

Inside, the contractions couple through the amniotic fluid to squeeze the fetus in a tight, liquid grip. Under these hydraulic stresses, fetal vessels compress, reducing blood flow to fetal tissues. The fetus responds to these forces with a heart rate that slows with each uterine contraction and a compensatory increase or rebound in heart rate after each contraction [27]. These rebounds in heart rate with each uterine contraction are primary signs of a healthy fetal vascular response and are believed by some to indicate that autonomic control of the heart is complete.

The fetal heart begins beating at about 170 beats per minute and slows to about 140 beats per minute near the fortieth week of gestation. The vascular system with its *foramen ovale* and *ductus arteriosus* is ready for the transfer. The rebounding heart rate, which may exceed 200 beats per minute with each contraction, indicates that controls for the cardiovascular system are ready. But until the first attempts to breathe air on the outside, the condition of the lungs remains unknown. Everything seems ready as birth begins.

The birth process, called labor, is divided into two time segments. The division is based on the amount of uterine dilation. The first stage of labor occurs when the cervix and uterus undergo a fetal-head-induced dilation. This dilation usually lasts from 8 to 24 hours. The second stage of labor begins once the cervical dilation is complete and involves delivering the fetus through the birth canal. This second stage can be quite short but typically lasts up to 30 minutes.

In about 95 percent of biths, labor ends with the baby born head first. Any other fetal orientation is termed a *breech presentation* (although "breech" really refers to the buttocks). Breech presentations cause problems for both the mother and the fetus. A breech presentation that cannot be resolved usually means delivery by means of a surgical procedure that opens the lateral uterine wall, a *caesarian section*. At the moment the fetus enters our world, the placenta is usually still attached to the uterine wall, still supplying oxygen to the fetus, now neonate. So long as the placenta is still attached, the neonate does not have to breathe through its lungs, but at the first attempt to draw air into still collapsed lungs, adaptation begins.

Outside the uterus, the first fetal task is to secure respiratory independence. The placenta may still be attached to the uterine wall, but the umbilical arteries are closing, and the blood exchange between the newborn and the placenta is fast shrinking. Now oxygen must come through the lungs, and the first breath will change the mechanical architecture of the lungs. It is one of the most difficult breaths ever taken by the neonate. The energy expended in these first few breaths will be 10 to 15 times greater than the energy required for normal breathing [28]. The respiratory channels must be mechanically pulled open to expand the alveoli. If they fail to stay open after the first breath, however, the great effort expended for the first breath must be repeated on each subsequent breath. It is a physical exertion that will exhaust a newborn very quickly. For most newborns, however, the lungs are ready and unfold into a delicate network of tissue.

Getting that first breath is made more difficult by the fetal anatomy. Neonatal tissues are soft and compliant, and the newborn chest is shaped more like a cone than a cylinder, with the lower ribs folded like the ribs of an umbrella. The ribs are still relatively soft and provide poor rigidity for the straining muscles trying to expand the lungs. Moreover, these soft bones and springy tissue make it difficult to keep airways open when the lungs expire. As bones become more rigid and muscles gain strength, the newborn settles into an easier breathing pattern.

These first breaths after birth cue the subsequent changes in the circulation pattern. The first changes are in the lungs. As the airways expand with air, the small arteries and veins that course through the lung tissue also expand. This increase in vascular size decreases the vascular resistance dramatically, and blood flow through the lungs becomes eight to ten times higher than before. Pressure in the right atrium falls below the left atrium, and the flap valve on the *foramen ovale* closes, following this pressure difference. And as the lungs bring higher oxygen levels to the blood and tissues, the sharply rising oxygen tension in the blood signals the *ductus arteriosus* to close, and layers of muscle in the vessel contract, clamping it shut. With increasing oxygen comes a decreasing carbon dioxide content that also signals ductus closure, along with a pH moving toward normal. Mixed in is a hormone called *bradykinin* which stimulates closure of the duct.

Figure 6–3 shows the new pattern of blood flow. The placenta is gone, but despite an organization that looks adult, it still has two of the special fetal channels. If for some reason these channels should reopen, the neonate is in trouble. For most, the change is complete without a backward step, becoming a solid adult pattern with no major problems.

Because the neonate must maintain an adult pattern with still-functional fetal channels, the vascular pattern can teeter in either direction. The success of this transition depends upon many subtle factors, one of which is the condition of the lungs. And in turn, the final condition of the lungs depends upon the dynamics of the neonatal chest wall and the secretion of a key pulmonary substance called *surfactant*.

Along with the cardiovascular changes, a neonate warmed by its mother's body *in utero* must now generate its own heat. The body geometry is poor, however, for the baby's size offers a large skin surface area and a small internal volume. This is an ideal situation for rapid heat loss. In the wrong situation, a baby can easily lose too much body heat and become *hypothermic.* Hypothermia, in turn, leads to a chain reaction process in which a lowered body temperature depresses the very metabolic processes used to generate heat. Below a certain internal temperature, an infant will be unable to heat itself, and the infant can lose ground in a hurry unless it is given external help to restore a normal body temperature.

We take for granted the obvious need to keep a neonate warm and protected. But just how far can temperature regulation of the body be pushed? We have, fortunately, some indications from early explorations of the tip of South America. During the 1800s, the Indians living on the tip of South America, along the Straits of Magellan, lived as fishermen. This is not unusual except that they fished with their whole body submerged in sea water, summer and winter. They gathered shellfish from the shallows throughout the year, wearing no clothing. They started conditioning babies to such extremes in body temperature by rolling them in snow immediately after birth [29]. If a baby survived that treatment, submerging in 34-degree water as an adult must have seemed easy. The story has an unusual twist, however. These people lost their ability to withstand cold when they took up modern technology and, perhaps more importantly, a modern diet. Photographs taken during the late 1800s of these phenomenal people fishing submerged in a winter ocean provide an outstanding example of human adaptation.

Although gestation and birth move in an ordered sequence with few, if any, major problems for most living humans, many birth processes do have problems. The problems not only show some of the finer shades of physiology but also show some of the power of current medical therapy.

SURVIVAL: MALADAPTION AND DISEASE

Even today, changing from a fetal to an extrauterine circulation pattern is an insurmountable barrier to nearly 25,000 newborns each year, and they perish [30]. Most of these babies were born prematurely. Many had heart defects so profound or so critical to life that they died before anything could be done. And some had new forms of birth defects of the brain, heart, or circulation produced by drugs or other agents ingested by the mother during gestation. The price for failing to make the transition to the world outside the uterus can be high, leaving life-long debilitation or death in its wake.

Many of the survival problems facing the newborn or neonate stem directly from the birth process. They are really the consequences of events in progress, and not acquired disease. How then can a physician make a prediction for survival of a newborn? One answer is the Apgar Score, an evaluation performed by the delivering physician, first systematized by Dr. Virginia Apgar. The newborn's respiratory effort, heart rate, skin color, tone, and response to stimuli are weighed and assigned a value [31]. A low Apgar Score bodes ill for an infant struggling to make the transition. The estimate on survival focuses on two important organs: the brain and the heart.

Deprived of the placenta, the neonate must rely on its lungs to acquire oxygen. If the function of the placenta and the lungs are both compromised during birth, the loss of oxygen can put the newborn life at risk, exposing the neonate to brain damage, if the baby lives through the ordeal.

One way of cutting off both supplies of oxygen (placenta and lungs) is a breech presentation. When the infant is delivered feet first, its head remains inside the uterus during labor, and the baby cannot breathe with its lungs, leaving the placenta as the only remaining source of oxygen. At the same time, uterine contractions can compress the umbilical cord and limit the blood returning from the placenta, and in turn, oxygen to the neonate's brain.

Another potential source of problems is myconium, the fetal feces. They can be excreted by the fetus during labor, and inhaled with the first breath. If the newborn inhales only a small amount of myconium, it may end up with only mild distress and pneumonia. A lot of myconium, however, can lodge in the trachea, closing off air to the lungs completely. If air flow into the lungs is blocked too long, the result can again be brain damage.

With the first breath, the lungs open and oxygen penetrates to some 30 million alveoli. The neonatal alveolar capacity is about one-tenth the adult capacity, and the newborn will grow about 270 million more when the lung is mature [30]. This one-to-ten growth factor does not hold for other structures in the lung, however. Trachea, bronchii, and broncheoles are reduced in size but are still large enough for good air flow in the neonate. If the pathways were too small, air resistance would be too large for a newborn trying to breathe. The trachea, for example, is one-half the size of the adult trachea, and even this reduction increases the air resistance some five to six times. The lung structures are scaled in the neonate to fit within a small chest but are still large enough for easy air flow.

Within the lungs, the architecture of the alveoli clusters remains unchanged until maturity. Gripping each alveolar cluster is a dense network of capillaries and lymphatic channels unequalled in other organs. The remaining lung tissue is made of two distinct cell types, identified as Type I and Type II. Most of the lung's mechanical support is provided by Type I cells, while the Type II cells play a chemical role. Because gases must move between the blood and the alveolar spaces through a thin tissue layer, the fluids between the blood and the air within the alveoli influence pulmonary function.

The physical geometry of the alveolar surfaces and their small size creates an unusual set of forces from the fluid lining the alveolar surface. These forces tend to close the alveolar space and come from the fluid surface tension [28]. With no other help, a newborn would have to overcome these surface forces on each breath and reopen the alveoli that were closed on exhalation. At the same time, the small amount of fluid lining the lung is necessary to prevent lung dehydration. The body can reduce the surface tension of the fluid by secreting a substance called *surfactant*. This special molecule sits at the water surface, separating the water molecules, reducing the forces among the water molecules at the air-water interface. Molecules that break up water-surface tension are called detergents, so this surface-acting agent in the lung is a biological "detergent." Surfactant reduces surface tension in the alveolar fluids enough to prevent lung collapse in exhalation.

When a neonate enters the world with no surfactant, a pink *hyaline membrane* begins to cover the lining of the lung, and each time the infant exhales, water tension closes the lungs almost completely. At the same time, plasma leaks out of the lung tissue into the alveolar space, filling portions of the air space with fluid [29]. This additional fluid within the lung spaces reduces oxygen diffusion into the body, provided the neonate can even keep its lungs open. In this situation, the neonate requires enormous energy to breathe, and the respiratory muscles are soon exhausted. This disease, caused by a lack of surfactant, is named for the pink hyaline membrane covering the lung: hyaline membrane disease.

Why is the surfactant absent? Perhaps the most accurate answer to that question is prematurity of the lung. The lungs simply are not yet ready to function. Where does the surfactant come from? Surfactant appears when the Type II cells reach maturity. This suggests that the Type II cells secrete this essential molecule. But a mature Type II cell seems to secrete surfactant on a hormone cue: adrenal steroids.

A fetus under stress from labor and birth secretes steroids that in turn stimulate secretion of surfactant, readying the fetus for the outside world [30]. This dependence upon fetal secretion of steroids means that fetal adrenal glands must be mature and working to stimulate surfactant secretion. This knowledge has helped obstetricians prepare a premature fetus for early delivery by administering steroids to it *in utero*, or by delaying delivery until fetal steroid levels climb to proper values. Two simultaneous influences must be present for a working neonatal lung: Type II cell maturity and steroid hormones.

A number of identifiable conditions seem to increase the chances of hyaline membrane disease in a neonate. For example, the possibility of hyaline membrane disease increases if the mother is a diabetic, if the baby is premature, or if the baby is delivered by caesarian section [32]. The probability of disease also increases with a maternal history of membrane disease in previous pregnancies, with birth asphyxia (no oxygen), or birth hypoxia (very low oxygen), with fetal acidosis or hypothermia, or should the neonate be born the second of twins. With these clues, an attending physician can anticipate hyaline membrane disease and initiate necessary therapy.

But even if the physician should know that the disease will occur, no specific replacement therapy for hyaline membrane disease exists, except for supportive therapy to get the neonate past this period of immaturity until its lungs begin secreting surfactant. Steroid therapy can hasten lung development, but without some sort of supportive therapy the fetus with hyaline membrane disease cannot maintain the muscular energy requirements and will soon become too exhausted to breathe.

Even with surfactant secretion, a neonate arrives with lungs filled with fluid. To carry on normal pulmonary function, the lungs must clear out this fluid. Labor and movement through the birth canal squeeze out some fluid. Most of the remaining fluid resides in the trachea and a few bronchii. When the lungs expand and pulmonary blood flow sharply increases, fluid absorption in the lungs also increases, and much of the fluid simply moves through the lung tissue into the neonatal bloodsteam. In addition, the pulmonary lymphatic system moves a large amount of fluid back into circulation. Fluids in the respiratory channels can rapidly compromise lung function as part of pneumonia and pulmonary edema in adults. The pulmonary lymphatic system prevents fluid aggregation and keeps fluids to a minimum when they do appear for both the neonate and the adult.

Neonatal and adult blood flow patterns are shown in Figure 6–3 with the special ducts that remain intact in the neonate. These ducts must remain closed. Because closure of these ducts, especially the *ductus arteriosus*, depends upon the oxygen levels in the blood, any condition that decreases blood oxygen can make the circulation pattern unstable. When the *ductus arteriosus* closes, the portion of the cardiac output that bypassed the lung through the duct (about 80 percent) now flows through the lung vasculature to receive oxygen. If the oxygen tension is reduced through some mechanism such as hyaline membrane disease or pneumonia, or even an upper respiratory infection that decreases the neonate's ability to breathe through its nose (the neonate does not yet know how to breathe through its mouth; it is an obligate nasal breather), the duct could reopen, again shunting blood past the lungs. As blood bypasses the lungs, the system oxygen levels decrease, and the duct opens even more. A vicious circle can develop, sweeping the infant into major respiratory distress. The response involves more than the *ductus arteriosus*. Lung vasculature begins to close as carbon dioxide builds in the blood, in turn increasing right heart pressure, which reopens the *foramen ovale*, shunting even more blood past the lungs. Untreated, this sort of vascular conversion to a fetal circulatory pattern can quickly produce shock and death.

Moreover, the effects of the compromised circulation extend to other systems. For example, as the vascular pattern reverses, low oxygen delivery to the tissues reduces the tissue's ability to generate heat. With a small body size that promotes heat loss, hypothermia can follow right on the heels of such a vascular change. And detecting hypothermia is not easy, because a newborn has no shivering mechanism. During this period, body heat comes from normal cellular metabolism and a special tissue called brown fat, which has a very high metabolic rate [33]. This sort of fat tissue is also found in hibernating animals and contributes to restoring an animal's core temperature during the waking process. The neonate has only a limited amount of this tissue, so this fat is not a large source of heat, but it does add to the total development of body heat. As such, brown fat is a rather unusual fat, with an ability to consume oxygen some 20 times faster than white fat, a rate that even exceeds the metabolic level of hard-working heart muscle [33].

Threats to body temperature are easy to anticipate, but an incomplete fetal immune system is hard to measure before an infection. Except for a few viruses that can penetrate to the interior of the uterus, the neonate enters the world with an uneducated immune system. In the womb, all fetal nutrients travel to the placenta through the maternal circulation, past an educated immune system that possesses both a memory of past infections and the maturity to respond to new ones. The fetus gets some protection from maternal antibodies that can pass through the placenta [6]. At birth, however, this protective maternal shield is gone. At birth, the microorganisms residing in the birth canal are the first exposure for the newborn. Nursing during the early neonatal period, however, provides an infant with many of the maternal antibodies through mother's milk, expressing the mother's disease experience [34]. These antibodies remain in the newborn, helping the separation from the maternal immune system.

Even if a newborn has a functioning set of lungs and does not fall prey to hypothermia, it still faces potential problems with its heart. About 32,000 infants are born each year with congenital heart defects [35]. Untreated, 50 to 60 percent would die within a year, 30 percent within a month. Often the problems are not apparent until well after birth.

If the heart malformation is an immediate threat, the attending physician must make a prompt evaluation of the neonatal condition, using techniques such as the Apgar Score and echocardiography.

The therapies available to handle congenital heart defects fall into two categories: curative and palliative. Some defects of the heart require a major rebuilding of the heart's anatomy. These defects are often so life-threatening that the first step is palliative repair to keep the infant alive long enough to plan the next surgical step. Other defects are debilitating in the long term and give the surgeon time to understand the defect and plan the surgery.

Before surgery, however, a congenital heart defect can cause two major forms of cardiovascular distress. The most frequent is congestive heart failure—that is, a heart severely overworked by the heart defect (see Chapter 3 for symptoms). The second killer is *hypoxemia*, a condition in which blood oxygen levels are too low to keep the body's cells alive. Frequently, the neonate's body responds to this chronically low oxygen level by making more red blood cells. The increased number of red blood cells makes the blood more viscous and prone to clotting. Tissue infarctions in the heart and brain are not uncommon byproducts of this pathologically increased hematocrit (the percentage of the blood volume occupied by red cells) [30]. Heart failure and blood clots may seem to be more adult problems than infant problems, but they really represent the way the cardiovascular system fails regardless of age.

Congestive heart failure can result from abnormal communications between pulmonary and systemic circulations. Such communications often occur at either the ventricular level or in the great vessels (pulmonary artery, pulmonary vein, and aorta). When lung vasculature increases in size, and its vascular resistance decreases, the pulmonary pressure can still be equal to or greater than the aortic pressure before the change in pulmonary circulation is complete. When pulmonary blood pressure falls below aortic pressure, the flow defect becomes evident. Now, the left side of the heart must carry not only the systemic volume load, but also the shunt through the defect. This creates a volume overload for the left heart as well as an increase in blood volume for the pulmonary circulation. An unclosed *ductus arteriosus* is the most frequent cause of this sort of volume overload.

But an open fetal duct is only one way in which flow through the vascular pattern can be abnormal. Venous flow from the lungs can be obstructed, causing blood to back up into the right heart. Alternatively, flow from the left ventricle can be obstructed, leading to an increase in pressure load for the left ventricle. This produces cardiac hypertrophy and eventually ventricular failure. If the obstruction is severe and the heart is failing, cardiac output will be low, causing a low systemic oxygen level in the bargain.

Systemic hypoxemia in the newborn can also result from two basic types of problems: First, blood flow to the lungs can be directly obstructed, preventing blood oxygenation; and second, the great vessels can be transposed, preventing any blood from reaching the lungs at birth because the lung circulation and the systemic circulation become independent, closed circulations. Figure 6–4 shows the results of exchanging the pulmonary artery with the aorta. The series organization of the heart's chambers is lost, forming two independent flow patterns. The only way blood from the pulmonary circulation can mix with systemic flow is through an open atrial or ventricular septal defect. The first surgical step is often to surgically enlarge the *foramen ovale*, providing an opening for mixing pulmonary and systemic blood. This usually offers enough mixing to keep the infant alive, but not much more. Later, surgery can more completely repair the series organization of the heart and restore normal tissue oxygenation.

Even when the heart and lungs of a newborn are working properly, it can be threatened by other life forms. Without an immune system capable of warding off the intrusion of a bacterium or a virus, an invading organism enters a warm, moist environment inside the

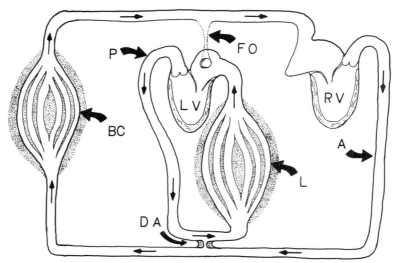

Figure 6.4. *Transposition of the great vessels. A transposition (exchange) of the great vessels creates two independent circulations, where the left heart (LV) pumps through the lungs only (L), and the right heart (RV) pumps through the body circulation only (BC). The abnormal pattern attaches the aorta (A) to the right ventricle and the pulmonary artery (PA) to the left ventricle. A patent ductus arteriosus (DA) can keep the neonate alive through limited blood mixing until corrective surgery.*

infant. One of the most virulent bacteria to infect the neonate is *Strep-tococcus B* [31]. The strep organism lives in the mother's birth canal without causing any problems for her. Infections from this organism can swiftly take an infant's life within 2 to 12 hours after birth. The disease strikes about 15,000 yearly, almost as many as were taken by polio during the 1950s [31].

This organism can be carried on and in an adult body with no overt signs of infection. When Strep B invades an adult body for the first time, a potential mother may experience a low-grade fever for a short time; then the infection settles into obscurity, going undiagnosed and unnoticed [31]. At birth, the neonate is exposed to the organism, and if the neonate is particularly vulnerable to infection and becomes infected, the result can be blindness, deafness, mental retardation, or death.

Immunization might be a way of beating this sort of infection. Research continues on a vaccine that would shift the emphasis to preventive medicine by immunizing women of childbearing age against this particular organism. Much is known, but much remains to be learned about this problem.

Thus far, we have been discussing problems in which the cause is apparent. But death can sometimes appear suddenly in the new-born, leaving no obvious evidence of the cause. This phenomenon is called sudden infant death syndrome (SIDS). Nearly 10,000 newborns die this way each year [36].

Sudden infant death syndrome is, as the name indicates, sudden, with no early indications of things to come. Events start with a runny nose or a slight cold, most often in winter or spring. The baby has no fever, no chills. The victim is usually two to four months old. Clues to the cause of death are scant, leaving more questions than answers.

Some research results suggest a subtle yet complex learning prob-lem on the part of the newborn. If the learning stage fails, the sleeping infant simply stops breathing and dies. At birth, a newborn is supplied with a set of instinctual or automatic reactions to blockage of the respiratory passages, or even simple head restraint [36]. As brain matu-ration and growth continue after birth, many instinctual responses are traded for learned responses; that is, the infant shifts control of breathing over to a more conscious level of the brain. If the "learning" process does not take place while the early automatic controls recede, the neonate will not make it past the learning stage. Some common findings in sudden infant death babies suggest this hypothesis to be very close to true.

One measure of a neonate's condition, as pointed out earlier, is the Apgar Score. This score supplies a number that represents the operating level of the brain and heart. SIDS babies have generally low

Apgar Scores, suggesting some mild brain dysfunction at birth. Along with a low Apgar Score there are often other abnormalities, such as poor physical response to physical head restraint or air passage blockage, poor feeding habits, and poor temperature regulation. And complicating events is the newborn's requirement that it breathe through its nose. (Mouth breathing, however, could make nursing risky for an infant trying to breathe without inhaling milk. Obligate nose breathing could be a natural defense mechanism for early neonatal life.)

After considering all factors, we may find we are not dealing with a single cause. The result may be the same stemming from several different causes. An alternate hypothesis is also gaining ground. Some infant deaths may result from a form of infant botulism, where the deadly spores are grown inside the infant's intestinal tract, poisoning the infant and leaving no easy way for investigators to find evidence [37].

CONCLUSIONS

The journey from conception to birth is probably the most compelling at its beginning and end, the moment of conception and the physical adaptation of birth. And a single cell starts the journey.

A long-existing question in this whole process is how the programming for tissues and organs occurs. The small yet well-defined anatomy of the cell is just now starting to explain how this can happen. We can expect to hear a great deal more about this process of differentiation in the future.

During the gestational period, the fetus takes on several classical roles: tissue allograft, nutritional parasite, and hypoxia adept. During this period, the two immune systems remain alert and separate; hormones from the conceptus and maternal system support the continued pregnancy; and through mechanics and biochemistry the fetus grows and develops at a surprisingly low oxygen level.

Like the road between two cities that is remembered for its five great bumps and not for the remaining smoothness, it is tempting to view birth as a disease rather than a natural process, considering only the dangers. But our discussion of dangers was only to show the fine line of survival each newborn must progress along. Yet for all the complexity involved, events that can take an infant's life do not come about that frequently. Healthy babies abound, and many of those who are stricken are healed by the present level of medical care.

As the national birth rate drops slowly, and human crowding limits the number of infants that can be accommodated in the world, each birth takes on a singular importance. It is a long, complex journey from the single cell formed by the fusion of sperm and ovum to the helpless creature cast out from its small, protective sea into our care. Some will not survive the process, but those who do should be treated with the special care we give any traveler making an arduous and momentous journey, one we all have taken, the most grand of all tours.

REFERENCES

Chapter 1

1. Quate, Calvin F., "The Acoustic Microscope," *Scientific American*, 241, no. 4 (October 1979), 62–70.

2. Lehninger, Albert L., *Biochemistry, The Molecular Basis of Cell Structure and Function*, p.43. New York: Worth Publishers, Inc., 1970.

3. Ibid., Ch. 2.

4. Davenport, Horace W., *Physiology of the Digestive Tract*, Ch. 17. Chicago: Year Book Medical Publishers, Inc., 1966.

5. Brown, G. H., J. W. Doane and V. D. Neff, *A Review of the Structure and Physical Properties of Liquid Crystals*, p. 26. Cleveland: CRC Press, 1971.

6. Katz, Bernard, *Nerve, Muscle, and Synapse*, p. 38. New York: McGraw-Hill Book Company, 1966.

7. Dowben, Robert M., *General Physiology, a Molecular Approach*, Ch. 7, "Cell Membranes." New York: Harper and Row, 1969.

8. Cereijido, Marcelino, and Catalina A. Rotunno, *Introduction to the Study of Biological Membranes*, pp. 140–45. New York: Gordon and Breach Science Publishers, 1970.

9. Katz, Bernard, *Nerve, Muscle, and Synapse*, p. 59.

10. Burnet, F. M., *Immunology, Aging, and Cancer*, Ch. 4. San Francisco: W. H. Freeman and Co., 1976.

11. Miller, Julie Ann, "The Bone and Muscle of Cells," *Science News*, 112 (October 15, 1977) 250–53.

12. Diem, K., and C. Lentner, eds., *Scientific Tables*, 7th ed., p. 523. Basle, Switzerland: Ciba-Geigy Limited, 1970.

13. Duke, James H., and John C. Bowen, "Fluids and Electrolytes: Basic Concepts and Recent Developments," *Contemporary OB/GYN,* 8 (October 1976), 177–94.

14. Goodman, L. S., and A. Gilman, *The Pharmacological Basis of Therapeutics,* 4th ed., p. 1210. New York: The Macmillan Co., 1970.

15. Lehninger, Albert L., *Biochemistry, The Molecular Basis of Cell Structure and Function,* pp. 396-99.

16. Ibid., Ch. 16.

17. Ibid., pp. 459-64.

18. Ibid., Ch. 14.

19. Lehninger, Albert L., *Bioenergetics,* pp. 75–77. New York: W. A. Benjamin, Inc., 1965.

20. Porter, Ian H., *Heredity and Disease,* pp. 1–4. New York: McGraw-Hill Book Company, 1968.

21. Benzer, Seymour, "The Fine Structure of the Gene," in *The Molecular Basis of Life.* San Francisco: W. H. Freeman and Co., 1968.

22. Porter, Ian H., *Heredity and Disease,* p. 44.

23. Gurdon, J. B., R. A. Lasky, and O. R. Reeves, "The Development Capacity of Nuclei Transplanted from Keratinized Skin Cells of Adult Frogs," *J. Embryol. Exp. Morph.,* 34 (1975), 93–112.

24. Lehninger, Albert L., *Biochemistry, The Molecular Basis of Cell Structure and Function,* Ch. 28.

25. Goss, Richard J., "Hypertrophy versus Hyperplasia," *Science,* 153 (September 1966), 1615–20.

26. Reitsma, W., "Skeletal Muscle Hypertrophy after Heavy Exercise in Rats with Surgically Reduced Muscle Function," *Amer. J. Phys. Medicine,* 48, no. 5 (1969), 237–58.

27. Jensen, David, *The Principles of Physiology,* pp. 114–16. New York: Appleton-Century-Crofts, 1976.

28. Dowben, Robert M., *General Physiology, a Molecular Approach,* pp. 545–48.

29. Bloom, William, and Don W. Fawcett, *A Textbook of Histology* (9th ed.), pp. 60-69. Philadelphia: W. B. Saunders Co., 1968.

30. Dowben, Robert M., *General Physiology, a Molecular Approach,* p. 558.

31. Bloom, William, and Don W. Fawcett, *A Textbook of Histology* (9th ed.) p. 49.

32. Mazia, Daniel, "The Cell Cycle," *Scientific American,* 230, no. 16 (January 1974), 54–64.

33. Prescott, David M., "Biology of Cancer and the Cancer Cell: Normal and Abnormal Regulation of Cell Reproduction," *CA: A Cancer Journal for Clinicians,* 22, no. 4 (July/Aug 1972), 262–72.

34. Bloom, William, and Don W. Fawcett, *A Textbook of Histology* (9th ed.), p. 54.

35. Moore, Keith L., *Before We Are Born,* p. 46. Philadelphia: W. B. Saunders Co., 1974.

36. Dowben, Robert M., *General Physiology, A Molecular Approach,* pp. 331–33.

37. Cairns, John, *Cancer: Science and Society*, p. 22. San Francisco: W. H. Freeman and Company, 1978.

38. Cairns, John, "The Cancer Problem," *Scientific American*, 233, no. 4 (November 1975), 64–78.

39. "A Graphic Stress Telethermometry System for Detection of Breast Disease," *Radiology/Nuclear Medicine Magazine* (April 1980), 1–3.

40. Folkman, Judah, "The Vascularization of Tumors," *Scientific American*, 234, no. 5 (May 1976), 59–73.

41. Salmon, Sydney E., "Malignant Disorders," in *Current Medical Diagnosis and Treatment 1983*, Marcus A. Krupp and Milton J. Chatton, eds. Los Altos, CA: Lange Medical Publications, 1983.

42. Baltimore, David, "Retroviruses and Cancer," *Hospital Practice* (January 1978), 49–57.

43. Cairns, John, *Cancer: Science and Society*, Ch. 4, "The Epidemiology of Cancer." San Francisco: W. H. Freeman and Company, 1978.

44. Cairns, John, *Cancer: Science and Society*, p. 149.

45. Burnet, F. M., *Immunology, Aging, and Cancer*, p. 139.

46. Trump, B. F., and Antti U. Arstila, "Cellular Reaction of Injury," in *Principles of Pathobiology* (2nd ed.) eds. Mariano F. La Via and Rolla B. Hill, Jr. New York: Oxford University Press, 1975.

47. Ibid., p. 11.

48. Ibid., p. 19.

Chapter 2

1. Bloom, William, and Don W. Fawcett, *A Textbook of Histology* (9th ed.) p. 272. Philadelphia; W. B. Saunders Company, 1968.

2. Ibid., p. 274.

3. Lehninger, Albert L., *Biochemistry, the Molecular Basis of Cell Structure and Function*, p. 590. New York: Worth Publishers, Inc., 1970.

4. Murray, John M., and Annemarie Weber, "The Cooperative Action of Muscle Proteins," *Scientific American*, 230, no. 2 (February 1974), 59–71.

5. Dowben, Robert M., *General Physiology, a Molecular Approach*, pp. 522–25. New York: Harper and Row, 1969.

6. Lehninger, Albert L., *Biochemistry, the Molecular Basis of Cell Structure and Function*, p. 290.

7. Ibid., pp. 590–91.

8. Ibid., p. 313.

9. Ibid., p. 316.

10. Weber, Annemarie, and John M. Murray, "Molecular Control Mechanisms in Muscle Contractions," *Physiological Reviews*, 53, no. 3 (July 1973), 613–73.

11. Vander, Arthur J., James H. Sherman, and Dorothy S. Luciano, *Human Physiology, the Mechanisms of Body Function* (2nd ed.) p. 207. New York: McGraw-Hill Book Company, 1975.

12. Weber, Annemarie, and John M. Murray, "Molecular Control Mechanisms in Muscle Contractions," 614.

13. Vander, Arthur J., James H. Sherman, and Dorothy S. Luciano, *Human Physiology, the Mechanisms of Body Function* (2nd ed.), p. 201.

14. Hoyle, Graham, "How is Muscle Turned On and Off?," *Scientific American*, 222, no. 4 (April 1970), 85–93.

15. Porter, Keith R., and Clara Franzini-Armstrong, "The Sarcoplasmic Reticulum," *Scientific American*, 212, no. 17 (March 1965), 73–81.

16. Cohen, Carolyn, "The Protein Switch of Muscle Contraction," 233, no. 5 (November 1975), 36–45.

17. Katz, Bernard, *Nerve, Muscle, and Synapse*, p. 43. New York: McGraw-Hill Book Company, 1966.

18. Zierler, Kenneth L., "Mechanism of Muscle Contraction and Its Energetics," in *Medical Physiology* (12th ed.), ed. Vernon B. Mountcastle, p. 1157. St. Louis: C. V. Mosby Co., 1968.

19. Ibid., p. 1134.

20. Jensen, David, *The Principles of Physiology*, p. 126. New York: Appleton-Century-Crofts, 1976.

21. Cheek, D. B., *Human Growth*, pp. 326–37. Philadelphia; Lea and Febiger, 1968.

22. Cheek, D. B., and D. E. Hill, "Muscle and Liver Cell Growth: The Role of Hormones and Nutritional Factors," *Fed. Proc.*, 29 (1970), 1503–9.

23. Reitsma, W., "Skeletal Muscle Hypertrophy after Heavy Exercise in Rats with Surgically Reduced Muscle Function," *Amer. J. Physiol. Med.*, 48 (1969), 237–58.

24. Vander, Arthur J., James H. Sherman, and Dorothy S. Luciano, *Human Physiology, the Mechanisms of Body Function*, p. 216.

25. Jeffress, R. N., and J. B. Peter, "Adaptation of Skeletal Muscle to Overloading: A Review," *Bull. Los Angeles Neurol. Soc.*, 35 (1970), 134–44.

26. Welt, Louis G., and William B. Blythe, "Cations: Calcium, Magnesium, Barium, Lithium, and Ammonium," in *The Pharmacological Basis of Therapeutics* (4th ed.), eds. Louis S. Goodman and Alfred Gilman, p. 807. New York: The MacMillan Co., 1970.

27. Grossman, Moses, and Ernest Jawetz, "Infectious Diseases: Viral and Rickettsial," in *Current Medical Diagnosis and Treatment 1983*, eds. Marcus A. Krupp and Milton J. Chatton, p. 820. Los Altos, CA: Lange Medical Publications, 1983.

Chapter 3

1. Florey, Ernst, *An Introduction to General and Comparative Animal Physiology*, pp. 178–83. Philadelphia: W. B. Saunders Company, 1966.

2. Lehninger, Albert L., *Biochemistry, the Molecular Basis of Cell Structure and Function*, p. 183. New York: Worth Publishers, Inc., 1970.

3. Burton, Alan C., *Physiology and Biophysics of the Circulation*, p. 63. Chicago: Year Book Medical Publishers, Inc., 1965.

4. Ibid., p. 64.

5. Bloom, William, and Don W. Fawcett, *A Textbook of Histology* (9th ed.), p. 374. Philadelphia: W. B. Saunders Company, 1968.

6. Jensen, David, *The Principles of Physiology*, p. 645. New York: Appleton-Century-Crofts, 1976.

7. Bloom, William, and Don W. Fawcett, *A Textbook of Histology*, p. 363.

8. Burton, Alan C., *Physiology and Biophysics of the Circulation*, p. 175.

9. Wood, J. Edwin, "The Venous System," *Scientific American*, 213, no. 1 (January 1968), 86–96.

10. Jensen, David, *The Principles of Physiology*, p. 646.

11. Bard, Philip, "Regulation of the Systemic Circulation," in *Medical Physiology* (12th ed.), ed. Vernon B. Mountcastle, p. 193. St. Louis: The C. V. Mosby Company, 1968.

12. Ibid., p. 204.

13. Russell, Diane H., Kathleen T. Shiverick, Burt B. Hamrell, and Norman R. Alpert, "Polyamine Synthesis During Initial Phases of Stress-induced Cardiac Hypertrophy," *Amer. Journal Physiology*, 221, no. 5 (November 1971), 1287–91.

14. Bergstrom, Sune, "Prostaglandins: Members of a New Hormonal System," *Science*, 157 (28 July 1967), 382–91.

15. Cole, James S., "A Historical Review of the Concept of the Left Ventricle as a Pump," *Catheterization and Cardiovascular Diagnosis*, 3 (1977), 155–70.

16. Luisada, Aldo A., and Pachalla K. Baht, "Auscultation and Phonocardiography as Aids to Cardiac Diagnosis–II: Mechanism of the Heart Sound," *Practical Cardiology* (April 1977), 41–51.

17. Marshall, Jean M., "The Heart," in *Medical Physiology* (12th ed.), vol. 1, ed. Vernon B. Mountcastle, pp. 35–68. St. Louis: The C. V. Mosby Company, 1968.

18. Bloom, William, and Don W. Fawcett, *A Textbook of Histology*, p. 376.

19. Milnor, William R., "Blood Supply of Special Regions," in *Medical Physiology*, ed. Vernon B. Mountcastle, pp. 221–43.

20. Goss, Richard J., "Adaptive Growth of the Heart," in *Cardiac Hypertrophy*, ed. Norman R. Alpert, pp. 1-10. New York: Academic Press, 1971.

21. Korecky, B., and M. Beznak, "Effect of Thyroxine on Growth and Function of Cardiac Muscle," in *Cardiac Hypertrophy*, ed. Norman R. Alpert, pp. 55-64.

22. Powis, Raymond L., "Growth Hormone in Cardiac Hypertrophy Induced by Nephrogenous Hypertension," in *Recent Advances in Studies on Cardiac Structures and Metabolism*, vol. 8, ed. Paul-Emile Roy and Peter Harris pp. 413–25. Baltimore: University Park Press, 1975.

23. Greipp, Randall B., Edward P. Stinson, and Norman E. Shumway, "Heart," in *Transplantation*, ed. John S. Najarian and Richard L. Simmons, p. 555. Philadelphia: Lea and Febiger, 1972.

24. Cronin, Michael P., "Jogging and Cardiovascular Health," *CVP* (March/April 1977), 21–40.

25. Netter, Frank H., *Heart*, ed. Fredrick F. Yonkman, p. 87. Summit, NJ: Ciba-Geigy Corporation, 1969.

26. Ibid., p. 88.

27. Lown, Bernard, "Intensive Heart Care," *Scientific American*, 219, no. 1 (July 1968), 19–27.

28. Ross, Russell, and John Glomset, "Atherosclerosis and the Arterial Smooth Muscle Cell," *Science*, 180 (21 June 1973), 1332–39.

29. Fields, William S., "Aortocranial Occlusive Vascular Disease," *Ciba Clinical Symposia*, vol. 26. no. 4 (1974).

30. Benditt, Earl P., "The Origin of Atherosclerosis," *Scientific American*, 236, no. 2 (February 1977), 74–85.

31. Baltimore, David, "Retroviruses and Cancer," *Hospital Practice* (January 1978), 49–57.

32. Marx, Jean L., "Atherosclerosis: The Cholesterol Connection," *Science*, 194 (12 November 1976), 711–14.

33. Clifford, Peter C., Robert Skidmore, John P. Woodcock, et al., "Arterial Grafts Imaged Using Doppler and Real-Time Ultrasound," *Vascular Diagnosis and Surgery* (February/March 1981), 43–57.

34. Sokolow, Maurice, "Heart and Great Vessels," in *Current Medical Diagnosis and Treatment 1983*, pp. 188–98. Los Altos, CA: Lange Medical Publications, 1983.

35. Marshall, Jean M., "The Heart," in *Medical Physiology* (12th ed.), ed. Vernon B. Mountcastle, p. 54.

36. Chusid, Joseph G., "Nervous System," in *Current Medical Diagnosis and Treatment 1983*, p. 570. Los Altos, CA: Lange Medical Publications.

37. Moe, Gordon K., and Alfred E. Farah, "Digitalis and Allied Cardiac Glycosides," in *The Pharmacological Basis of Therapeutics* (4th ed.), ed. Louis S. Goodman and Alfred Gilman, p. 680. New York: The Macmillan Company, 1970.

38. Ibid., p. 684.

39. Moe, Gordon K., and J. A. Abildskov, "Antiarrhythmic Drugs," in *The Pharmacological Basis of Therapeutics*, (4th ed.), p. 722.

40. Burton, Alan C., *Physiology and Biophysics of the Circulation*, p. 187.

Chapter 4

1. Sperry, R. W., "Lateral Specialization of Cerebral Function in the Surgically Separated Hemispheres," in *The Psychophysiology of Thinking*, ed. F. J. McGuigan and R. A. Schoonover, pp. 209–29. New York: Academic Press, 1973.

2. Nocenti, Mero R., "Pituitary Gland," in *Medical Physiology*, (12th ed.), ed. Vernon B. Mountcastle, p. 871. St. Louis: The C. V. Mosby Company, 1968.

3. Halsey, James H., Jr., and Scott McFarland, "Oxygen Cycles and Metabolic Autoregulation," *Stroke*, 5 (March–April 1974), 219–25.

4. Maugh, Thomas H., II, "Diabetes: Epidemiology Suggests a Viral Connection," *Science*, 188 (25 April 1975), 547–51.

5. _____, "Hormone Receptors: New Clues to the Cause of Diabetes," *Science*, 193 (16 July 1976), 220–22.

6. Ganong, William F., *Review of Medical Physiology* (4th ed.), p. 49. Los Altos, CA: Lange Medical Publications, 1969.

7. Harper, Harold A., *Review of Physiological Chemistry* (12th ed.), p. 303. Los Altos, CA: Lange Medical Publications, 1969.

8. Kolata, Gina Bari, "Blood Sugar and the Complications of Diabetes," *Science*, 203 (16 March 1979), 1098–99.

9. Katz, Bernard, *Nerve, Muscle, and Synapse*, p. 153. New York: McGraw-Hill Book Company, 1966.

10. Moore, Keith L., *Before We Are Born*, p. 209. Philadelphia: W. B. Saunders Company, 1974.

11. Lehninger, Albert L., *Biochemistry, the Molecular Basis of Cell Structure and Function*, p. 413. New York: Worth Publishers, Inc., 1970.

12. Camargo, Carlos A., and Felix O. Kolb, "Endocrine Disorders," in *Current Medical Diagnosis and Treatment 1983*, ed. Marcus A. Krupp and Milton J. Chatton, p. 684. Los Altos, CA: Lange Medical Publications, 1983.

13. Schally, Andrew V., Akira Arimura, and Abba J. Kastin, "Hypothalamic Regulatory Hormones," *Science*, 179 (26 January 1973), 341–50.

14. Gillie, R. Bruce, "Endemic Goiter," *Scientific American*, 224, no. 17 (June 1971), 93–101.

15. Jensen, David, *The Principles of Physiology*, p. 1099. New York: Appleton-Century-Crofts, 1976.

16. Ibid., p. 1100.

17. Ibid., p. 1097.

18. Tepperman, J., *Metabolic and Endocrine Physiology* (2nd ed.), p. 115. Chicago: Year Book Medical Publishers, Inc., 1968.

19. Ibid., p. 128.

20. Pitts, Robert F., *Physiology of the Kidney and Body Fluids* (2nd ed.), p. 153. Chicago: Year Book Medical Publishers, Inc., 1968.

21. Ibid., p. 156.

22. Sutherland, Earl W., "Studies on the Mechanism of Hormone Action," *Science*, 177 (4 August 1972), 401–8.

23. Pastan, Ira, "Cyclic AMP," *Scientific American*, 227, no. 13 (August 1972), 97–105.

24. O'Malley, Burt W., and Anthony R. Means, "Female Steroid Hormones and Target Cell Nuclei," *Science*, 183 (15 February 1974), 610–20.

25. DiCara, Leo V., "Learning in the Autonomic Nervous System," *Scientific American* 222, no. 16 (January 1970), 31–39.

26. Eyzaguirre, Carlos, *Physiology of the Nervous System*, pp. 85–91. Chicago: Year Book Medical Publishers, Inc., 1969.

27. Skinner, B. F., *Beyond Freedom and Dignity*. New York: Bantam Books, 1972.

28. Houk, James, and Elwood Henneman, "Feedback Control of Movement and Posture," in *Medical Physiology* (12th ed.), ed. Vernon B. Mountcastle, p. 1688. St. Louis: The C. V. Mosby Company, 1968.

29. Lippold, Olaf, "Physiological Tremor," *Scientific American*, 224, no. 14 (March 1971), 65–73.

30. Henneman, Elwood, "Peripheral Mechanisms Involved in the Control of Muscle," in *Medical Physiology* (12th ed.), p. 1707.

31. Chusid, Joseph G., and Joseph J. McDonald, *Correlative Neuroanatomy and Functional Neurobiology* (13th ed.), "Reflexes," pp. 208–13. Los Altos, CA: Lange Medical Publications, 1967.

32. Bard, Philip, "Regulation of the Systemic Circulation," in *Medical Physiology* (12th ed.), p. 204.

33. Netter, Frank H., *Heart*, vol. 5, ed. Fredrick F. Yonkman, p. 18. Summit, NJ: Ciba-Geigy Corporation, 1969.

Chapter 5

1. Bloom, William, and Don W. Fawcett, *A Textbook of Histology* (9th ed.), p. 501. Philadelphia: W. B. Saunders Company, 1968.

2. Ibid., p. 497.

3. Ibid., p. 499.

4. Ibid., p. 812.

5. Smith, Alice Lorraine, *Principles of Microbiology* (8th ed.), p. 52. St. Louis: The C. V. Mosby Co., 1977.

6. Ibid., p. 430.

7. Grossman, Moses, and Ernest Jawetz, "Infectious Diseases: Bacterial," in *Current Medical Diagnosis and Treatment 1983*, ed. Marcus A. Krupp and Milton J. Chatton, p. 837. Los Altos, CA: Lange Medical Publications, 1983.

8. Marples, Mary J., "Life on the Human Skin," *Scientific American*, 220, no. 18 (January 1969), 108–15.

9. Halde, Carlyn, "Infectious Diseases: Mycotic," in *Current Medical Diagnosis and Treatment 1983*, p. 927.

10. Vander, Arthur J., James H. Sherman, and Dorothy S. Luciano, *Human Physiology, the Mechanisms of Body Function* (2nd ed.), p. 320. New York: McGraw-Hill Book Company, 1970.

11. Bloom, William, and Don W. Fawcett, *A Textbook of Histology* (9th ed.), p. 561.

12. Lambertsen, Christian J., "The Atmosphere and Gas Exchanges with the Lungs and Blood," in *Medical Physiology* (12th ed.), ed. Vernon B. Mountcastle, p. 647. St. Louis: The C. V. Mosby Company, 1968.

13. Ganong, William F., *Review of Medical Physiology* (4th ed.), p. 544. Los Altos, CA: Lange Medical Publications, 1969.

14. *Carbon Monoxide*, p. 80, Committee on Medical and Biological Effects of Environmental Pollutants, National Academy of Sciences, Washington, D.C., 1977.

15. Wynder, Ernest L., and Dietrich Hoffmann, "Certain Constituents of Tobacco Products," Ch. VIII, *Tobacco and Tobacco Smoke*. New York: Academic Press, 1967.

16. Lyons, Albert S., and R. Joseph Petrucelli II, *Medicine, an Illustrated History*, p. 90. New York: Harry A. Abrams, Inc., 1978.

17. Furst, Peter T., and Michael D. Coe, "Ritual Enemas," *Natural History*, 36, no. 3 (March 1977), 88-91.

18. Davenport, Horace W., *Physiology of the Digestive Tract* (2nd ed.), p. 216. Chicago: Year Book Medical Publishers, Inc., 1966.

19. Ibid., p. 217.

20. Hirschorn, Norbert, and William B. Greenough III, "Cholera," *Scientific American*, 225, no. 2 (August 1971), 15–21.

21. Cizek, Louis J., "The Kidney," in *Medical Physiology* (12th ed.), p. 309.

Chapter 6

1. Kolata, Gina, "Studying Learning in the Womb," *Science* 225 (20 July 1984), pp. 302–03.

2. Moore, Keith L., *Before We Are Born*. Philadelphia: W. B. Saunders Co., 1977, p. 20.

3. Lyon, Mary F., "Gene Action in the X Chromosome of the Mouse (Mus Musculus L.)," *Nature* 190 (1961), p. 372.

4. Kolata, Gina Bari, "Developmental Biology: Where Is It Going?" *Science* 206 (19 October 1979), 315–16.

5. Miller, Julie Ann, "The Bone and Muscle of Cells," *Science News* 112 (October 15, 1977), 250–53.

6. Beer, Alan E., and R. E. Billingham, "The Embryo as a Transplant," *Scientific American*, 230, no. 16 (April 1974), pp. 36–41.

7. Moore, Keith L., *Before We Are Born*, p. 28.

8. Benson, Ralph C., "Gynecology and Obstetrics," in *Current Medical Diagnosis and Treatment 1983*, p. 481. Marcus A. Krupp and Milton J. Chatton, eds. Los Altos, CA: Lange Medical Publications, 1983.

9. Dawood, M. Yusoff, "Hormones in Amniotic Fluid," *Am. J. Obstet. Gynecol.* 128 (1977), pp. 576–83.

10. Jensen, David, *The Principles of Physiology*, p. 1197. New York: Appleton-Century-Crofts, 1976.

11. Ibid., p. 1198.

12. Ibid., p. 1199.

13. Harper, Harold A., and Grodsky, G. M., "The Chemistry and Function of Hormones," in *Review of Physiological Chemistry*, (12th ed.), p. 482. Los Altos, CA: Lange Medical Publications, 1969.

14. Ibid., pp. 476–77.

15. Hobbins, John C., and Fred Winsberg, *Ultrasonography in Obstetrics and Gynecology*, p. 88. Baltimore: The Williams and Wilkins Company, 1977.

16. Dawes, G. S., *Foetal and Neonatal Physiology*, p. 19. Chicago: Year Book Medical Publishers, Inc., 1968.

17. Ibid., p. 22.

18. Ibid., p. 23.

19. Assali, N. S., "Some Aspects of Fetal Life *in utero* and the Changes at Birth," *Amer. J. Obstet and Gynec.*, 97 (1 February 1967), p. 324.

20. Peltonen, T., and L. Hirvonen, "Experimental Studies on Fetal and Neonatal Circulation," *Acta. Paed. Scan. Suppl*, 161 (1965), pp. 1–55.

21. Assali, N. S., "The Fetus and the Neonate," in *Biology of Gestation*, Vol. II. New York: Academic Press, 1968.

22. Dawes, G. S., *Fetal and Neonatal Physiology*, p. 95.

23. Ibid., p. 97.

24. Fletcher, C. M., and Daniel Horn, in *WHO Chronicle*, Vol. 24 (1970). Geneva, Switzerland, pp. 345–70.

25. Jensen David, *The Principles of Physiology*, p. 1206.

26. Minei, Lawrence J., and Kataro Suzuki, "Role of Fetal Deglutition and Micturition in the Production and Turnover of Amniotic Fluid in the Monkey," *Obstetrics and Gynecology*, 48, no. 2 (August 1976), pp. 177–81.

27. Fox, Harold E., "Basics of Fetal Monitoring," *Perinatal Care* 2, no. 5 (May 1978), pp. 26–36.

28. Smith, Clement A., "The First Breath," *Scientific American* 209, no. 4 (October 1963), pp. 27–35.

29. *Peoples of the Earth*, Vol. 7, *The Andes*, ed. Peter Riviere, pp. 120–23. Verona, Italy: The Danbury Press, 1973.

30. Avery, Mary Ellen, Nai-San Wang, and H. William Taeusch, Jr., "The Lung of the Newborn Infant," *Scientific American*, 228, no. 11 (April 1973), pp. 75–85.

31. Gluck, Louis, "Special Problems of the Newborn," *Hospital Practice*, (January 1978), pp. 75–88.

32. Meier, Werner A., "Fetal Respiratory Development and Adaptation to Extrauterine Life," *Peri- and Neonatology* (September–October 1977), pp. 29-45.

33. Dawkins, Michael J. R., and David Hull, "The Production of Heat by Fat," *Scientific American* (August 1965), pp. 62–67.

34. Peterson, John C., and William Bock, "Educating Nursing Mothers," *Perinatal Care*, 2, no. 7 (July–August 1978), pp. 44–47.

35. Tyson, Kenneth R. T., "Congenital Heart Disease in Infants," *Clinical Symposia*, 27, no. 3 (1975), ed. by Robert J. Shapter, p. 2. Summit, NJ: Ciba Pharmaceutical Company, 1975.

36. Marx, Jean L., "Crib Death: Some Promising Leads but No Solution Yet," *Science* 189 (1 August 1975), 367–68.

37. _____, "Botulism in Infants: A Cause of Sudden Death?" *Science* 201 (1 September 1978), pp. 799–801.

Index

INDEX